Reactor Laboratory Experiments at Kyoto University Critical Assembly

Cheol Ho Pyeon · Go Chiba · Tomohiro Endo ·
Kenichi Watanabe

Reactor Laboratory Experiments at Kyoto University Critical Assembly

Springer

Cheol Ho Pyeon
Institute for Integrated Radiation
and Nuclear Science
Kyoto University
Osaka, Japan

Tomohiro Endo
Graduate School of Engineering
Nagoya University
Nagoya, Japan

Go Chiba
Graduate School of Engineering
Hokkaido University
Sapporo, Japan

Kenichi Watanabe
Graduate School of Engineering
Kyushu University
Fukuoka, Japan

ISBN 978-981-97-8069-3 ISBN 978-981-97-8070-9 (eBook)
https://doi.org/10.1007/978-981-97-8070-9

This work was supported by Ministry of Education, Culture, Sports, Science and Technology.

This Springer imprint is published by the registered company Springer Nature Singapore Pte Ltd.
The registered company address is: 152 Beach Road, #21-01/04 Gateway East, Singapore 189721, Singapore

Preface

The Kyoto University Critical Assembly (KUCA) was established by official approval of the Japanese government in 1972 as one of the reactor facilities in the Kyoto University Research Reactor Institute (KURRI) (currently, Institute for Integrated Radiation and Nuclear Science, Kyoto University: KURNS). The KURNS is one of the national joint-use research institutes and is open to researchers affiliated with universities and research institutes in Japan. In addition, KURNS is expected to play an important role as the research institute that provides many opportunities in conducting basic research of various fields related to nuclear energy and radiation applications, through the operation of large experimental facilities, including the Kyoto University Research Reactor (KUR; approved in 1962) and KUCA.

KUCA has three reactors (two solid-moderated and solid-reflected cores: A and B cores, and one light-water-moderated and light-water-reflected core: C core) in one building, which has an operating system with the selection of one core among multiple cores. At KUCA, basic research associated with reactor physics has been conducted mainly in the three reactors: in the A core, the development of subcriticality measurement methods and the feasibility of accelerator-driven system; in the B core, neutronics characteristics of thorium-fueled cores; and, in the C core, neutronics characteristics of tight-pitch-fueled cores and measurement methodologies of neutron spectrum. Through the experiments in the three cores, unique research results have been successfully produced by many researchers in domestic and overseas universities and research institutes.

In 1975, KUCA has launched an educational program called the Joint Reactor Laboratory Experiments, in the C core, for graduate students in Japan. This program was based on the "Laboratory Experiments" conducted at the Argonne National Laboratory, U.S., in the late 1960s (Daavettila DA et al. (1965) A manual of reactor laboratory experiments. ANL-6990, Argonne). Immediately after its start in 1975, a simplified version of the textbook was prepared by faculty members of Tohoku University, Tokyo Institute of Technology, Tokai University, Nagoya University, Osaka University, and Kyoto University for graduate students majoring in reactor physics and radiation detection. Basic experiments on reactor physics, including approach-to-criticality and control rod calibration experiments and neutron flux measurements, were conducted for graduate students at the six universities. The program has been established as an experimental education program jointly organized eleven universities in Japan, including Hokkaido University, Tohoku University, Nagaoka University of Technology, Tokyo Institute of Technology, Tokyo City University, Tokai University, Nagoya University, Kyoto University, Osaka University, University of Fukui, and Kyushu University. The program offers intermediate and advanced courses according to the undergraduate and graduate students' level of mastery of nuclear-related subjects. In the previous time, for undergraduate and graduate students from Korea (2003 through 2016) and Sweden (2006 through 2010), laboratory experiments had been conducted as nuclear human resource development programs in both countries. Since 2017, for graduate students majoring in reactor physics in Korea and China, the program has been offered at Seoul National University, Kyung Hee University, Hanyang University, Korea Advanced Institute of Science and Technology, Pohang University of Science and Technology, and Ulsan National Institute of Science and Technology in Korea; and Tsinghua University in China, as the Reactor Physics Asia Experiment Program (RPHA-XP: αXP).

The simplified textbook of the Joint Reactor Laboratory Experiments at KUCA had covered experimental topics, including approach-to-criticality and control rod calibration experiments, reaction rate measurement by activation foil method, and subcriticality measurements by Feynman-α method and pulsed neutron source method. After many revisions and modifications, the book was published in 2010 as *Nuclear Reactor Physics Experiments* (Misawa T, Unesaki H, Pyeon CH (2010) Kyoto University Press, Japan; http://hdl.handle.net/2433/276400) and was available to all participants in both domestic and overseas.

This book is an educational textbook on reactor physics experiments jointly written by Prof. Tomohiro Endo of Nagoya University, Prof. Kenichi Watanabe of Kyushu University, Prof. Go Chiba of Hokkaido University, and the author. The three professors, who are experts in reactor physics (Prof. Endo and Prof. Chiba) and radiation detection (Prof. Watanabe), have devoted much time and effort to the writing of this book. This book is based on a careful examination of the contents and educational benefits obtained from the previously published *Nuclear Reactor Physics Experiments*. Also, on the valuable lessons learned and deep reflections gained in previous educational programs, and the significant revisions to the previously published version of the textbook, we have tried to write a textbook that is easier to understand than ever before. In addition, the textbook has been updated to reflect the latest developments in radiation detection technology. Through digital processing of response data from detectors employed in radiation detection techniques ("Neutron Measurements," in Chap. 9), we have attempted to improve the accuracy and real-time measurement of reactor physics parameters. The latest findings and theories of reactor physics and radiation detection are developed in this book. For example, a part of Chap. 2, and all of Chaps. 4 and 5 of *Nuclear Reactor Physics Experiments* are merged into a new chapter (Chap. 10) entitled "Subcriticality Measurements." The inclusion of *Subcriticality Measurements* is one of the major features of this book. At KUCA, "Exponential Experiments with Natural Uranium and Measurement of Uranium Enrichment" (Chaps. 7 and 8, respectively) have been conducted for some students. By adding these two experimental topics to this book, most of the experimental topics required for reactor physics experiments are covered. Furthermore, the book includes comprehensive items on experiments using nuclear reactors, involving basic and applied topics on radiation detection. This book is expected to provide students participating in the experiments with an understanding of the basic knowledge about reactor physics parameters required for the operation of nuclear reactors through the reactor physics experiments.

This book is written for graduate students who have already studied reactor physics and radiation detection in their undergraduate courses. Readers who have not studied reactor physics and radiation detection as undergraduates or who wish to review them again should read the sister book, *Introduction to Nuclear Reactor Experiments* (Wakabayashi G, Yamada T, Endo T, Pyeon CH (2023) Springer, Singapore; Open Access, https://link.springer.com/book/10.1007/978-981-19-6589-0). The sister book contains basic concepts and theories of reactor physics and radiation detection and is a must-have for all readers majoring in nuclear engineering. The authors hope that this book, together with *Introduction to Reactor Experiments*, will provide a systematic understanding of reactor physics experiments.

This book consists of the following contents: Chap. 1 provides an overview of the KUCA-C core. Chapter 2 presents fundamental theories of reactor physics, and Chaps. 3 and 4 describe the basic and most important reactor physics experiments: approach-to-criticality and control rod calibration experiments, respectively. In Chaps. 5 and 6, detailed methodologies are described, involving the measurement of reaction rate and determination of neutron flux using the activation foil method, respectively. Chapters 7 and 8 indicate the contents of the exponential experiments with natural uranium and the measurement of uranium enrichment, respectively. Chapters 9 and 10 show the contents of digital processing of experimental data and various measurement methods for subcriticality, respectively. The measurement methods described in Chap. 9 are used to attain the subcriticality described in Chap. 10, and the contents could be somewhat difficult to understand, but do not give up and persevere in readers' efforts to understand the contents.

We are grateful to Prof. Emeritus Ikuo Kanno of Kyoto University, Prof. Akihiro Nohtomi of Kyushu University, and Dr. Hiroki Iwamoto of Japan Atomic Energy Agency for reviewing the chapters in which they specialize and providing valuable comments. We would like to take this opportunity to thank them.

The authors are thankful to Springer Nature for having the patience to complete the manuscript and their responses throughout its preparation for about a year.

Finally, it should be noted that the textbook was published with the subsidy of the project for "Establishment of Nuclear Education Platform Focusing on University Research Reactors and Large-Sized Facilities" of "Global Nuclear Human Resource Development Initiative 2024" in the Advanced Nuclear Education Consortium for the Future Society (ANEC) founded by the Ministry of Education, Culture, Sports, Science, and Technology (MEXT) in Japan.

Osaka, Japan
May 2024

Prof. Cheol Ho Pyeon

Contents

Chapter 1
Kyoto University Critical Assembly

Abstract The Kyoto University Critical Assembly (KUCA) has three types of cores: two solid-moderated and solid-reflected cores (A and B cores), and one light-water-moderated and light-water-reflected core (C core). The three cores are operated at room temperature and very low power less than 1 W, as a normal operation mode, and referred as zero-power reactors. Also, they have reactor components, including the low-enriched uranium fuel with the uranium-235 enrichment less than 20 wt%, moderators, reflectors, three control rods, three safety rods, six neutron detectors (three fission chambers and three uncompensated ionization chambers). Of three cores, in this chapter, essential parts of the C core are mainly introduced to make an easy understanding of the reactor core itself, including core characteristics, utilization purposes, functions of shutdown systems, and core components.

Keywords KUCA · Zero-power reactor · Low-enriched uranium fuel · Light-water moderator · Light-water reflector

1.1 Light-Water-Moderated and Light-Water-Reflected Core

The Kyoto University Critical Assembly (KUCA; Fig. 1.1) was installed with the main purpose of conducting basic research on neutronics characteristics associated with reactor physics. It is used for verifying the accuracy of calculation codes and nuclear data, and validating calculation models, as referred benchmark experiments. Also, KUCA is used for conducting nuclear education and training for undergraduate and graduate students in domestic and overseas universities through various reactor physics experiments. KUCA is one of the few multi-table experimental facilities in the world, and is located at the Institute for Integrated Radiation and Nuclear Science, Kyoto University (KURNS; former Kyoto University Research Reactor Institute, Kumatori-cho, Sennan-gun, Osaka, Japan). KUCA has three types of cores including A [1, 2], B, and C cores [3–5], and the users can select the operating core according to the purposes of the experiments. Of three cores, the A and B cores are

C. H. Pyeon et al., *Reactor Laboratory Experiments at Kyoto University Critical Assembly*, https://doi.org/10.1007/978-981-97-8070-9_1

Fig. 1.1 Photo of KUCA building. © KURNS. All rights reserved

termed the solid-moderated and solid-reflected cores that use polyethylene or graphite as moderator and reflector, and the C core is termed the light-water moderated and light-water-reflected core that uses light water as moderator and reflector.

In the C core, silicide-dispersed fuel (low-enriched uranium: LEU) consisting of less than 20 wt% enriched uranium (U), aluminum (Al), and silicon (Si): U_3Si_2-Al is used, and coated with Al, and loaded into an Al fuel element support frame (hereinafter termed fuel frame) to form the fuel assembly. Until 2020, U-Al alloy consisting of 93 wt% enriched uranium (highly-enriched uranium: HEU) [3–5] and Al was used as the fuel element.

The C core has a core tank made of aluminum with a diameter and depth of 2000 mm, and a stainless-steel lattice plate at the bottom of the tank. The core is composed of several fuel assemblies loaded on the lattice plate. A single core system can be assembled by loading several fuel assemblies, while a coupled core can be assembled by dividing the lattice plate supporting the fuel assemblies into two parts with arbitrary widths. In addition, control rods, safety rods, neutron source insertion tube, and various detectors are set in the core tank. The light water that is the moderator and reflector in the C core is supplied from the lower part of the core tank.

Figure 1.2 shows a photo of a core assembled in the C core. The fuel assemblies are loaded in the core tank, and the control rods, safety rods, and various detectors are arranged around the fuel assemblies.

Fig. 1.2 Photo of top view of C core © KURNS. All rights reserved

1.2 Characteristics

The maximum thermal power is 100 W, whereas the power during normal operation of the reactor is on the order of mW, less than 1 W, termed a zero-power reactor. Due to the very low power of the reactor, KUCA, like most critical assemblies, has the following characteristics;

(1) Since the fission product yield is roughly proportional to the reactor power, the accumulation of fission products is very low.
(2) The reactor core is exposed and has no concrete biological shielding. Radiation leakage from the core is shielded by the building walls.
(3) Fuel manipulation does not require remote control, and core reconfiguration is then easy (due to above items (1) and (2)).
(4) The low power of the reactor means that there is no accumulation of xenon and the reactor can be shut down at any time. Also, the reactor can be restarted quickly.

Because the power of the reactor is very low and, consequently, the radiation exposure is very low, researchers and experimenters tend to think that they can conduct various experiments with a high degree of freedom at their discretion, as shown in the above characteristics of the experiments carried out in criticality assemblies. When conducting experiments, however, it is required by law that the reactor core be configured in compliance with various restrictions and that the reactor be operated safely. To

Table 1.1 Reactor limits of light-water-moderated and light-water-reflected core at KUCA

Characteristic	Detail	Limits
Reactivity	Excess reactivity	Lower than 0.5 $\%\Delta k/k$
	Total control rod worth	More than 1.5 $\%\Delta k/k$ which is sum of maximum excess reactivity and 1 $\%\Delta k/k$
	Maximum control rod worth of one rod	Lower than one-third (1/3) of total reactivity (3 control rods and 3 safety rods)
	Reactivity of tank water	More than 1 $\%\Delta k/k$
Reactivity coefficient	Maximum differential reactivity	Lower than 2×10^{-2} $\%\Delta k/k \cdot s^{-1}$
	Temperature coefficient	Lower than 2×10^{-2} $\%\Delta k/k \cdot °C^{-1}$ at a near critical state
Thermal limit	Temperature of moderators and reflectors	Lower than 80 °C
Ratio of moderator to fuel	Atomic number ratio of $^{1.1}H/^{235}U$*	Lower than 400
Thermal power		100 W (at most)

$^{1.1}H/^{235}U$*: hydrogen-1/uranium-235

carry out experiments, the main reactor limits specified for the light-water-moderated and light-water-reflected core at KUCA are required, as shown in Table 1.1.

1.3 Utilization Purposes

KUCA, like most critical assemblies, has three main utilization purposes, which are described in this section.

1.3.1 Mockup Experiments

Usually, when designing a new type of reactor, the core design is performed using various calculation codes and nuclear data, but ultimately their validity should be examined by using critical assemblies. In many cases, the same procedure is also adopted when modifying the core. Experiments conducted to design such a new type of reactor are called mockup experiments. Numerical calculations using calculation codes and nuclear data do not always accurately determine integral parameters of the target core system and include errors. Therefore, to finally evaluate the accuracy of numerical calculations, it is effective to compare the results of numerical calculations with those of experiments obtained from critical assemblies, and the accuracy and reliability of numerical calculations can be assured from the comparison between experiments and calculations.

1.3.2 Benchmark Experiments

All physical quantities measured in critical assemblies are integral quantities. When reactor parameters, including critical mass, control rod worth, neutron flux distribution, and temperature coefficient, are obtained by numerical calculations, the integral quantity is obtained using neutron cross sections and through various processes. By comparing between the results of experiments and numerical calculations, the validity of the nuclear data, calculation methods, or calculation models used in the numerical calculations can be examined. Here, rather than the development of a new reactor, experiments using critical assemblies are sometimes conducted primarily for the purposes of examining the calculation methods or calculation models themselves. The experiments are called benchmark experiments, and are conducted using relatively simple geometries and compositions. Benchmark experiments are said to be the most suitable for the purposes of using critical assemblies because they are mainly conducted with the core in cold and clean conditions (cold: no fuel burnup, clean: no extra neutron-absorbing material in the core).

1.3.3 Experiments for Education

As described in Sect. 1.2, the characteristics of critical assemblies are that the amount of fission product yield that accumulates in the fuel is small, the radiation contained in the fuel and radiation leakage during operation is low, and the fuel can be directly handled. Also, the characteristics could provide easy access to the reactor core. KUCA can provide undergraduate and graduate students with opportunities to experience nuclear reactor experiments [3, 6] through many experiments that can be carried out in critical assemblies. Since 1975, nuclear education programs for undergraduate and graduate students in domestic universities have been conducted at KUCA as the Joint Reactor Laboratory Experiments, and educational programs for overseas undergraduate and graduate students have been managed since 2003.

1.4 Emergency Shutdown

1.4.1 Outline

In the light-water-moderated and light-water-reflected core, the procedure for immediately shutting down an operating reactor in the event of an emergency is as follows:

- Three control rods and three safety rods, which are suspended by the electromagnetic power supply using a control rod drive mechanism, are inserted by disconnecting the electromagnetic power supply (to increase the absorption effect of neutrons in the core).
- Dump valves are opened and emergency water level processing equipment is activated to drain tank water from the core (remove light-water moderator and light-water reflector from the core).

We now turn to a description of the devices and functions described as follows:

KUCA has three control rods and three safety rods as neutron absorbers. Control rods (C1, C2, and C3) are mainly used for control of criticality and power of the reactor. Safety rods (S4, S5, and S6) are always withdrawn to the upper limit during reactor operation (including loading of fuel assemblies), and in an emergency, all control rods and safety rods are inserted into the reactor core by free fall to shut down the reactor. The control rods and the safety rods are divided into an absorber (Al pipe wrapped with cadmium: Cd plates) and a control rod drive mechanism, both of which are connected by an electromagnet power supply. By turning off the electromagnetic power supply, the control rods and the safety rods can be inserted into the reactor core by natural fall due to gravity.

The emergency shutdown system of KUCA has two types: "scram" and "simultaneous insertion (auto run-down)." Here, simultaneous insertion occurs when the requirements for activation are slightly less stringent than for a scram, and is said to be a pre-scram before a scram event occurs. However, once the simultaneous insertion is triggered, the scram occurs instantly afterward, so it is positioned as a warning before the scram is triggered. The simultaneous insertion is said to be one of the emergency shutdown functions unique to KUCA, which is rarely seen in other reactors.

Although the origin of scram is not known, it is said that the world's first nuclear reactor (Chicago Pile No. 1: CP-1 [7]) was already equipped with an emergency shutdown function when it was built in the United States in December 1942. The function at that time was called "Sudden Control Rod Activation Mechanism," SCRAM. It is also said that the term SCRAM is derived from slang in the U.S. In any case, SCRAM has been established as a term for a function to shut down a nuclear reactor in an emergency, and is still used today in test and research reactors, and boiling water reactors.

In the light-water-moderated and light-water-reflected core at KUCA, SCRAM and simultaneous insertion are triggered and, in some cases, alarms are issued when the following events occur in the reactor shown in Sect. 1.4.2.

1.4.2 Alarm Operation

At KUCA, an alarm is triggered when one of the following events occurs:

- When the stable period (see Chap. 3) is lower than 30 s.
- When the core temperature exceeds 80 °C.
- When the water level in the furnace chamber pit exceeds 20 cm.
- When fires occur.

1.4.3 Pre-SCRAM

The events that activate simultaneous insertion are as follows:

- When the indication in each range of the linear power meter exceeds 110%.
- When the stable period is lower than 15 s.

When one of the above events occurs, the following functions work inside the reactor as a preliminary warning to immediately shut down the reactor: the three control rods remain connected to the control rod drive mechanism and electromagnetic power supply, and the control rods are automatically inserted at a constant speed by motor drive.

1.4.4 SCRAM

The events that activate the SCRAM are as follows:

- When the indication in each range of the linear power meter exceeds 120%.
- When the indication of the safety power meter exceeds 120%.
- When the stable period is lower than 10 s.
- When a malfunction occurs in the high-voltage power supply for nuclear instrumentation.
- When an error occurs in the power supply voltage in the control room.
- When the movable shielding door opens.
- When the track size door opens.
- When the overhead crane moves during the reactor operation.
- When the locking pins of the control rod drive mechanism fixing plate are removed.

- When there is an earthquake acceleration of more than 19.6 $cm\,s^{-2}$.
- When the emergency alarm button is pressed.
- When the manual SCRAM button is pressed.

When one of the above events occurs, the following functions work inside the reactor to immediately shut down the reactor:

- The electromagnetic power supplies for the control rods and the safety rods are disconnected, and a total of three control rods and three safety rods are immediately inserted into the core by natural fallout.
- By opening dump valves and activating the emergency water level lowering device, the light-water moderator and the light-water reflector are drained into the core tank and removed from the core.

1.5 Core Components

1.5.1 Fuel Plate

The fuel plate (Fig. 1.3) consists of U_3Si_2-Al (LEU) fuel meat (fuel material) in 0.5 mm thick coated with Al in 0.5 mm thick, with dimensions of 600 mm long, 62 mm wide, and 1.5 mm thick. The fuel meat contains less than 20 wt% enriched uranium.

1.5.2 Fuel Frame

In the C core, the fuel plates shown in Sect. 1.5.1 are inserted into an Al fuel frame to form the fuel assembly. The appearance of the fuel frame is shown in Fig. 1.4, the cross-sectional view (axial direction) is shown in Fig. 1.5, and the dimension relating to the loading of the fuel plates into the fuel frame is shown in Fig. 1.6 and Table 1.2.

As shown in Fig. 1.4, the fuel frame is 70 mm wide, 140 mm long, and 735 mm high (excluding the fixing legs). Three types of fuel frames are available for the C core, namely C30, C35, and C45 (Fig. 1.5 and Table 1.2). The smallest unit combining the fuel plate and the light water moderator region is called a unit cell (Fig. 1.6), and there are three different fuel array pitches. The fact means that there are three different lattices with a different atomic number ratio of $^{1}H/^{235}U$ contained per unit volume of the core. The C core allows experiments in which the neutron spectrum of the core is varied by selecting different lattices. Note that the number of fuel plates that can be loaded per fuel frame depends on the fuel arrangement pitch of that frame; therefore, in the C core, the number of fuel plates in a fuel frame can be changed in units of one plate according to a certain procedure, and the arrangement of fuel assemblies (core geometry) can also be changed. Furthermore, the water level

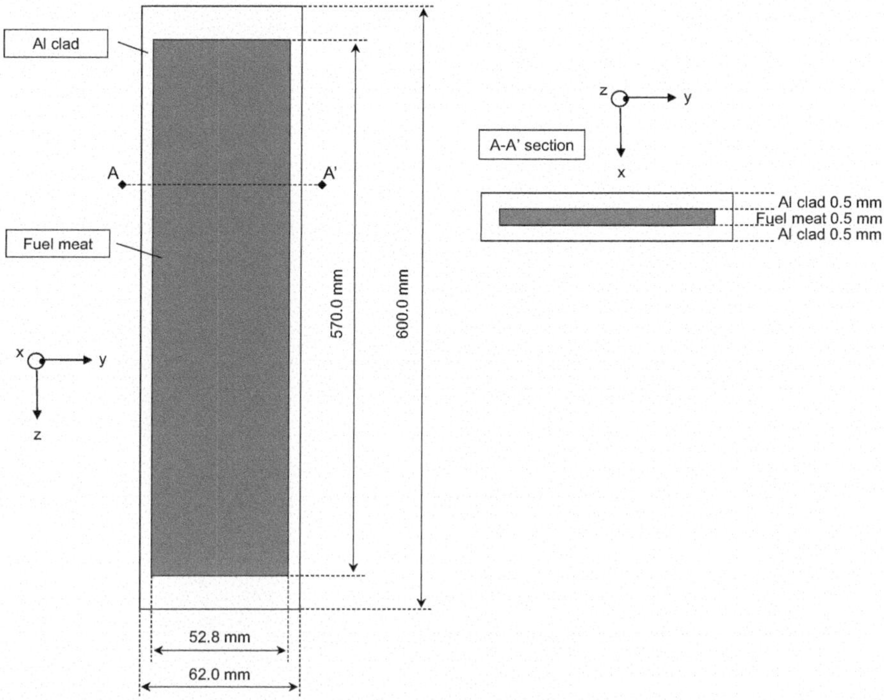

Fig. 1.3 Fuel plate used in C core

of the moderator can be changed, allowing various experiments to be carried out according to the purposes of the research or practical training.

Fuel frames are loaded into the core tank by standing on a grid plate. As shown in Fig. 1.7, there is a gap of 2 mm on the long side (direction x) and 1 mm on the short side (direction y) between adjacent fuel frames, which is also filled with light water during the operation. Therefore, when the core is divided into fuel frame units, the light-water gap should also be considered. When the core is considered as a repetition of a certain unit, the repetition unit is called a unit cell. In terms of the C core, the unit cell of fuel assembly is a combination of a fuel frame and a light-water gap per fuel frame, and the dimensions of the unit cell are 71 × 142 mm (Fig. 1.7). The various group constants used in criticality calculations should thus be obtained by considering the light-water gap for the unit cell of fuel assembly.

1.5.3 Control Rod and Safety Rod

In the C core, cylindrical Cd plates are used as neutron absorbers. The control rods and the safety rods are connected to the control rod driving mechanism by electromagnets.

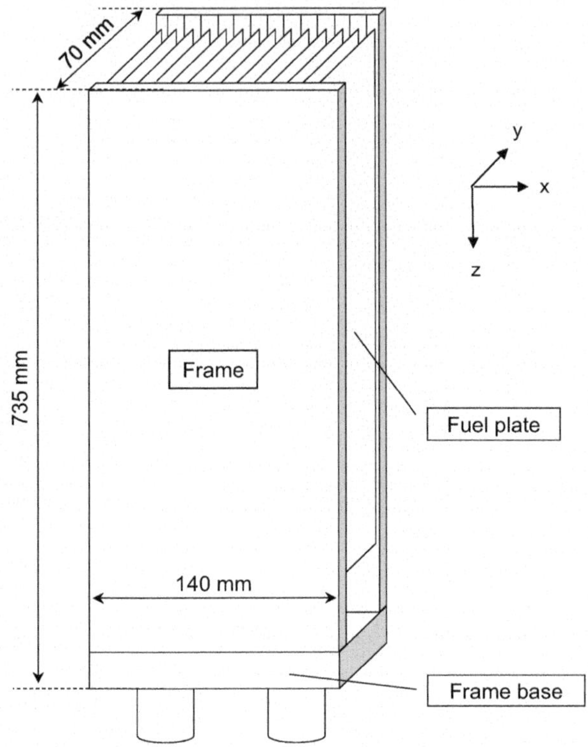

Fig. 1.4 External view of fuel frame of C core

In an emergency, the electromagnet current is cut off, causing the control rods and the safety rods to drop by gravity and the reactor to stop safely. The structure of control rods and safety rods is described in Chap. 4.

1.5.4 Neutron Source for Reactor Startup

The americium (Am)-beryllium (Be) neutron source (Am-Be: quantity 7.4×10^{10} Bq; neutron intensity $5 \times 10^{6}\ s^{-1}$; and energy 4 MeV) is used as the external neutron source required to start up the reactor. In the Am-Be neutron source, the neutrons are produced by $^{9}Be\ (\alpha,\ n)^{12}C$ reactions using α particles emitted from ^{241}Am. The capsule containing Am-Be has a diameter of about 20 mm and a length of about 50 mm. In experiments, the Am-Be neutron source in the capsule is treated as a point neutron source.

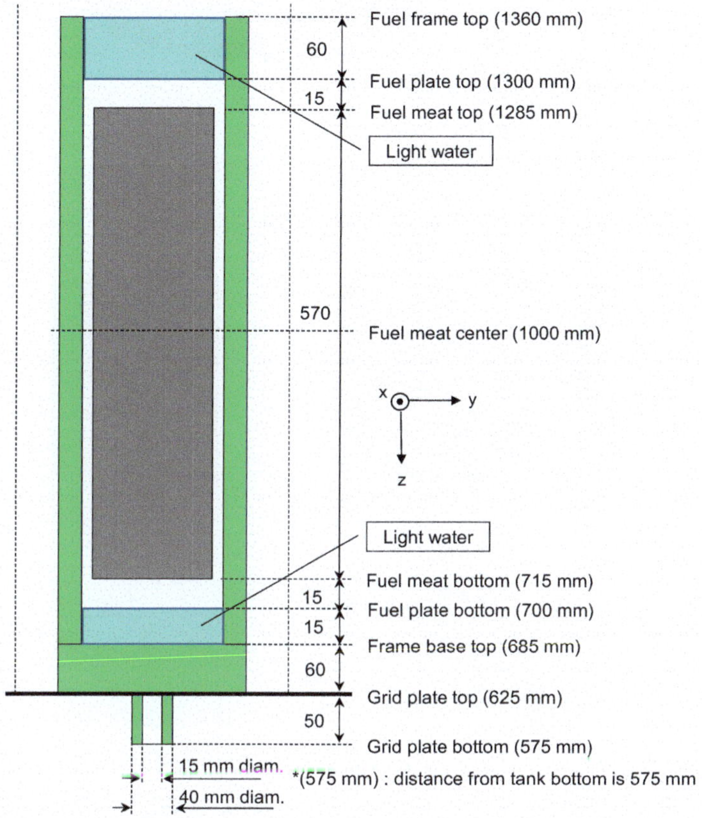

Fig. 1.5 Cross-sectional view of fuel frame of C core (direction z)

1.5.5 Neutron Detectors

There are six neutron measurement detectors on each of the KUCA cores, with three fission chambers (FCs) and three uncompensated ionization chambers (UICs) as neutron detectors. The FCs are called start-up systems and are mainly responsible for low-power region measurements during reactor start-up. The output is displayed as counting rates on a recorder at the front of the control desk, and can also be displayed digitally by means of a scaler. In reactors with high power, the startup system is used to monitor the power in the low-power region during reactor startup and may stop working when the reactor power increases as the reactor reaches a critical state. KUCA operates the startup system throughout all experiments because the reactor power is very low.

The three detectors using the UICs receive a higher output range than the FC and are called the logarithmic output system for stable period (UIC#4, Log-N), the linear output system for reactor power (UIC#5, Lin-N), and the safe output system

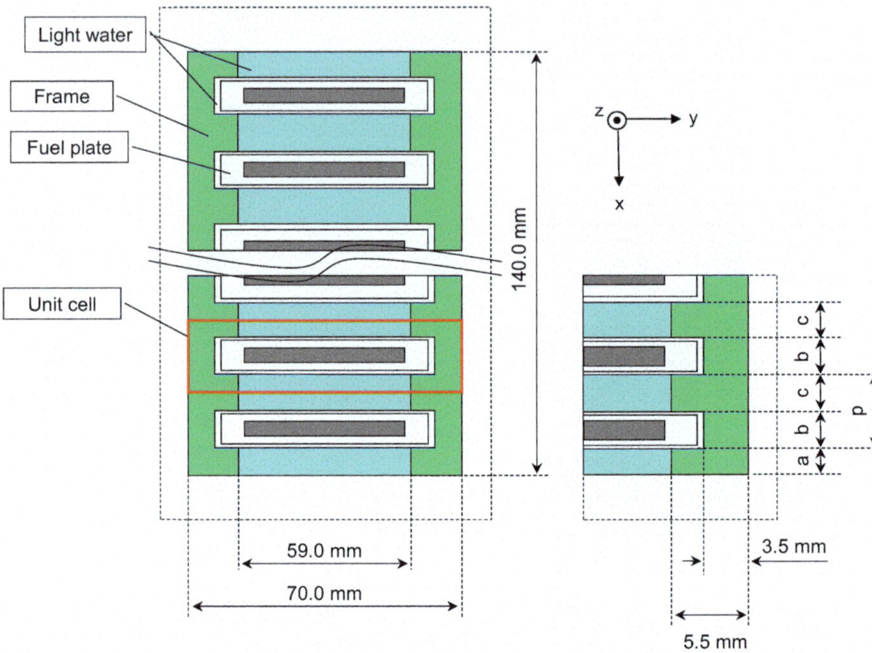

Fig. 1.6 Dimension of fuel frame of C core

Table 1.2 Detailed dimension of fuel frame of C core (see Fig. 1.6)

Frame	C30	C35	C45
a: frame edge (mm)	1.07	1.095	1.05
b: groove width (mm)	1.70	1.70	1.70
c: separation width (mm)	1.26	1.79	2.84
p: fuel pitch (mm)	2.96	3.49	4.54
Max. plates/frame	47	40	31

(UIC#6), depending on their purposes. Their outputs are recorded on recorders in front of the reactor console, and the outputs of the linear output system are displayed on digital indicators in the center of the reactor console. At KUCA, the signal level of the safety output system is extremely low because the reactor power is very low. Therefore, data from the safety output system are not used in experiments.

At KUCA, five measurement detectors, FC#1, FC#2, FC#3, UIC#4 and UIC#5 described above, are used to obtain experimental data. Depending on the purposes, several auxiliary detectors such as boron-trifluoride (BF_3) and helium-3 (^3He) neutron detectors could be set around the core for the Approach-to-criticality and control rod calibration experiments in Chaps. 3 and 4, respectively.

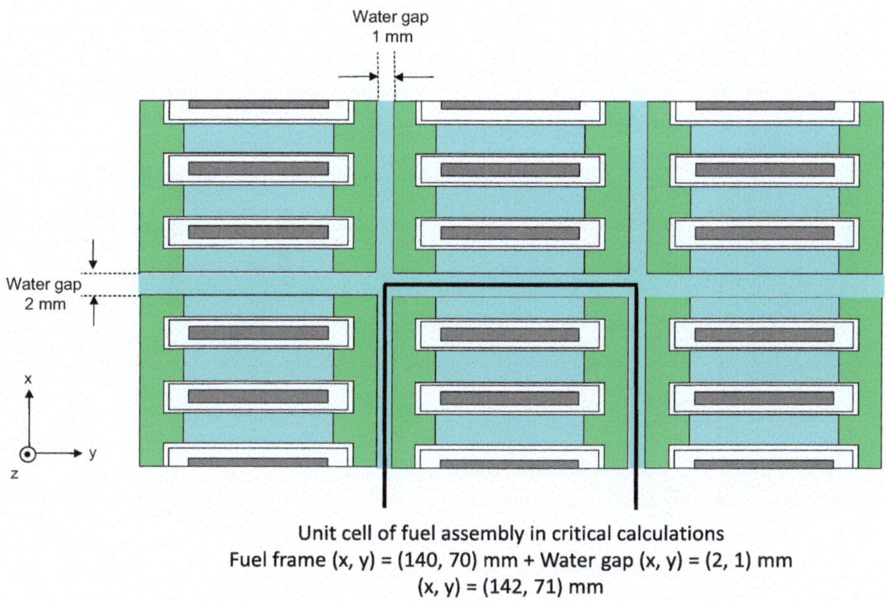

Fig. 1.7 Outline of unit cell of fuel assembly

1.5.6 Core Tank and Grid Plate

Fuel assemblies, control rods, safety rods, guide tube of neutron source, and various detectors are set on a stainless-steel grid plate at the bottom of the Al core tank. The grid plate has holes drilled in it for placing the fuel assemblies and other core components. The grid plate can also be divided into two parts of arbitrary width, and experiments on so-called coupled reactors can be performed by making two cores.

After the fuel assemblies and other components are loaded and the core components such as detectors are set, the reactor is finally ready to function when light water is supplied from the bottom of the core tank.

Figure 1.8 shows the relationship between the height positions of the grid plate, fuel assemblies (fuel frames and fuel plates), various detectors, and neutron source in the core tank, corresponding to values of water-level gauge by the indicator. It should be remembered that the value of water-level gauge, 1000 mm, corresponds to the center of the fuel plate meat.

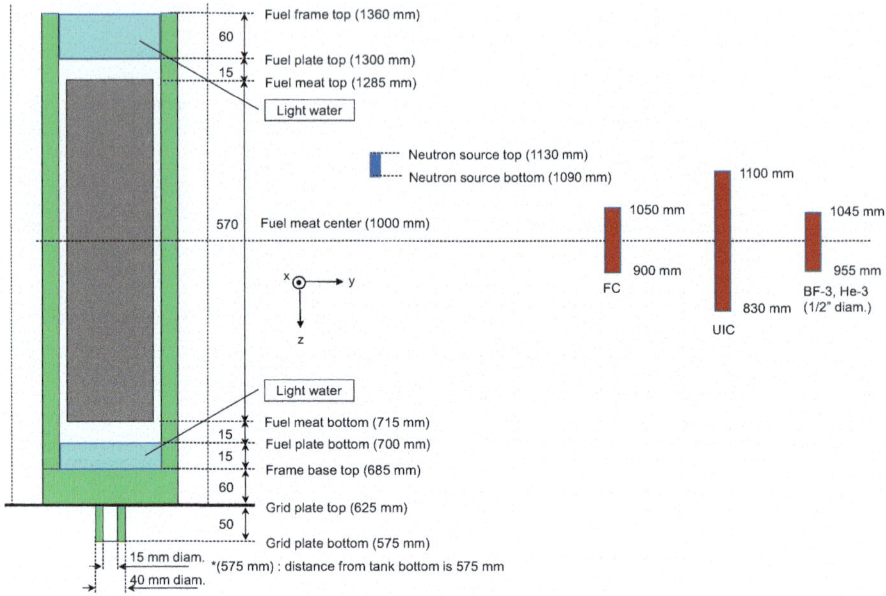

Fig. 1.8 Relationship between core components and water level of C core

1.5.7 Name of Experimental Core

To constitute the reactor, the steps are to fix the control rods, safety rods, guide tube of neutron source, and detectors on the grid plate in that order, insert fuel plates into fuel frames to make fuel assemblies, and load the fuel assemblies on the grid plate.

An example of a core configuration is shown in Fig. 1.9. When directions x and y are determined as shown in Fig. 1.9, in this example, 3 (direction x) $\times$ 5 (direction y) $=$ 15 fuel assemblies composed of C35 fuel frames are arranged. Also, it can be confirmed that control rods, safety rods, neutron source insertion tube, and detectors are set around fuel assemblies.

KUCA names the cores that are configured based on certain rules. For example, in terms of C core, the core name is determined as follows:

$$\text{C} + \text{lattice name} + \text{G} + \text{core gap(number of fuel assemblies in direction } y\text{)}$$

In the case of the core shown in Fig. 1.9, the core is given as the name C35G0(5) core because it has a 35 lattice, that is, 3.5 mm fuel arrangement pitch, G0 because the core is a single core with no gap (zero gap), and 5 because the fuel assemblies are arranged in five columns in the direction y.

There are three types of fuel lattices used in the experiments (C30, C35, and C45), and the number of columns in the fuel assembly arrangement depends on the purposes of the experiments. Special attention should be paid to the information on

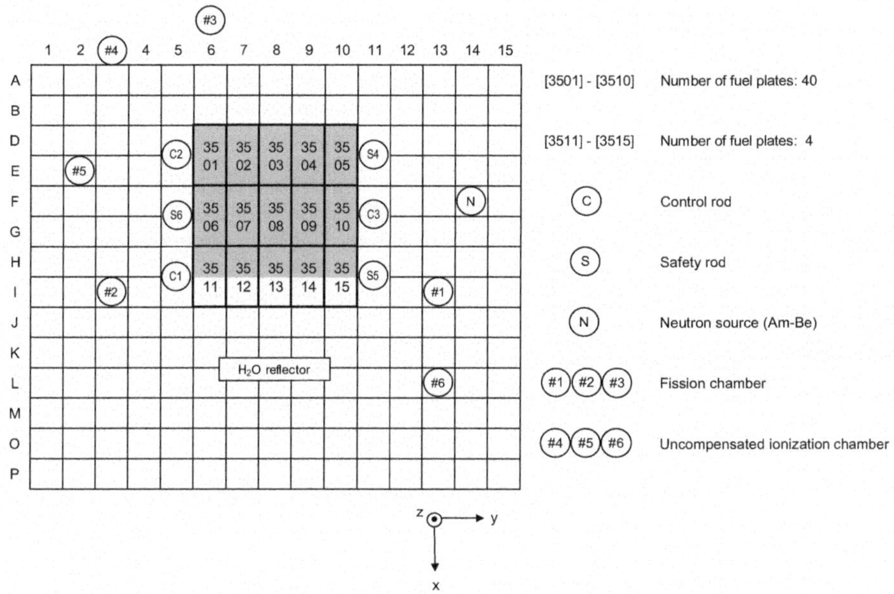

Fig. 1.9 Top view of configuration of C35G0(5) core (example with HEU fuel)

the configuration of the core, since the core configuration is an important parameter in numerical calculations of the prediction of the critical mass of fuel plates.

References

1. Pyeon CH ed (2021) Accelerator-driven system at Kyoto University Critical Assembly. Springer, Singapore https://doi.org/10.1007/978-981-16-0344-0. Accessed 24 May 2024
2. Pyeon CH, Morioka K (2022) Nuclear data-induced uncertainty of criticality in solid-moderated and solid-reflected cores with highly enriched uranium fuel at Kyoto University Critical Assembly. Nucl Sci Eng 196:1147–1160. https://doi.org/10.1080/00295639.2022.2070385
3. Misawa T, Unesaki H, Pyeon CH (2010) Nuclear reactor physics experiments. Kyoto University Press, Kyoto, Japan http://hdl.handle.net/2433/276400. Accessed 24 May 2024
4. Pyeon CH, Morioka K (2022) Monte Carlo analyses of light-water-moderated and light-water-reflected cores with highly-enriched uranium fuel at Kyoto University critical assembly. J Nucl Sci Technol 59:257–265. https://doi.org/10.1080/00223131.2021.1961636
5. Pyeon CH, Morioka K (2022) Uncertainty quantification of light-water-moderated and light-water-reflected cores with highly-enriched uranium fuel at Kyoto University critical assembly. J Nucl Sci Technol 59:898–906. https://doi.org/10.1080/00223131.2021.2017371
6. Wakabayashi G, Yamada T, Endo T, Pyeon CH (2023) Introduction to nuclear reactor experiments. Springer, Singapore. https://doi.org/10.1007/978-981-19-6589-0. Accessed 24 May 2024
7. U.S. Atomic Energy Commission, Division of Technical Information (1942) The fist reactor. Oak Ridge, Tennessee

Chapter 2
Basics of Nuclear Reactor Physics

Abstract When we conduct various kinds of experiments with nuclear reactors and interpret the obtained measurement results qualitatively and quantitatively, knowledge of the theory of nuclear reactor physics is quite helpful and it enables us to better understand the experiments. Phenomena occurring in nuclear reactors are generally based on the interections between neutrons and nuclei, so the fundamental knowledge on this interaction are described and several important physical quantities are introduced. Then, the fission chain reactions are interpreted by a classical theory, the six-factor formula. Based on the fundamental quantities in the nuclear reactor physics such as the neutron flux and the reaction rate, the neutron diffusion equation is derived, and the neutron non-leakage probabilities are quantified based on this diffusion theory. Furthermore, the basics of the nuclear reactor kinetics theory are discussed using the point kinetics equation.

Keywords Nuclear reactor physics · Interactions between neutron and nucleus · Reaction cross section · Neutron flux · Neutron diffusion equation · Nuclear reactor kinetics

2.1 Interactions Between Neutron and Nucleus

Nuclear reactors are systems in which nuclear fission chain reactions induced by neutrons sustainably occur under controls by mankind. This section describes the interactions between neutron and material (or nucleus in material), which are important to understand various physical processes occurring in nuclear reactors.

2.1.1 Scattering and Absorption Reactions

In this section, two main interaction types, scattering and absorption reactions, are explained.

C. H. Pyeon et al., *Reactor Laboratory Experiments at Kyoto University Critical Assembly*, https://doi.org/10.1007/978-981-97-8070-9_2

2.1.1.1 Scattering Reaction

Through the scattering reaction between neutron and nucleus, neutron is again emitted after the interaction. In the laboratory system, the emitted neutron moves with energy and direction which are different from those of the incident neutron. Scattering reaction can be further categorized into the elastic scattering and the inelastic scattering.

In the elastic scattering reaction, sums of kinetic energy and momentum of neutron and nucleus are preserved and the internal structure of the nucleus is unchanged through the reaction. The incident neutron generally loses its kinetic energy by transferring the fraction of its energy into the nucleus.

In the inelastic scattering, a fraction of the kinetic energy of the incident neutron is consumed to increase the internal energy of the target nucleus (or to make the target nucleus the excited state, or to change the internal structure of the target nucleus), and thus the sums of kinetic energy and momentum of neutron and nucleus are not preserved through the reaction.

While the number of emitted neutrons in these reactions is generally one, there are several other reactions through which multiple neutrons are emitted, and these reactions are also categorized into the scattering reaction. As an example, the (n, 2n) reaction, in which two neutrons are emitted after the interaction between neutron and nucleus, is well known.

2.1.1.2 Absorption Reaction

In the absorption reaction, neutron collides with and is absorbed by the nucleus, and a compound nucleus is formed. This compound nucleus then emits radiations and particles such as γ-ray and α-particle. Through this reaction, mass number of the target nucleus is increased by one and this nucleus, the compound nucleus, becomes in an excited state at first. The degree of the increase in the energy level is determined by the kinetic energy of the incident neutron and the neutron binding energy. The absorption reaction can be categorized into several reactions according to the types of the emitted radiations or particles.

Through the radiative capture reaction, the compound nucleus changes its state from the excited level to the ground level by emitting γ-ray. This reaction is denoted to as (n, γ), and sometimes is simply referred to as the capture reaction.

Through the charged-particle emission reaction, the compound nucleus emits charged-particles such as α-particle and proton. This reaction mainly occurs in light nuclides, and when the incident neutron kinetic energy is high, probability of this reaction becomes large in many nuclides. These reactions are denoted to as (n, α), (n, p), and so on.

In the nuclear fission reaction, the compound nucleus is divided into two fractions (fission fragments), and two or three neutrons on average are emitted at the same time. This reaction occurs mainly in heavy nuclides such as uranium (U) and plutonium (Pu). It is denoted to as (n, f). Neutrons emitted after this reaction can induce the next fission reactions, and thus nuclear fission reactions can sustainably occur without providing any neutrons from the outside of the system if certain conditions are satisfied. This is known as the fission chain reaction. The fission reaction has significant importance when the physics of nuclear reactors is concerned, so it is discussed in more detail in the following section.

2.1.2 Fission Reaction

In a nuclear reactor, there exist many neutrons with different energies in the range of 20 MeV to 10^{-5} eV.

In the case of ^{235}U, fission reaction can occur through the interaction with a low-energy neutron which is in an equilibrium with the ambient temperature of around several hundreds kelvin. Such a low-energy neutron is generally referred to as a thermal neutron. The energy of the thermal neutron corresponding to the room temperature of 293 K is 0.025 eV and its velocity is approximately 2200 m s^{-1}.

Through one fission reaction of ^{235}U induced by a thermal neutron, approximately 2.4 neutrons are emitted on average. Although the number of emitted neutrons is dependent on fission fragments at each fission reaction, its averaged value is important generally in the nuclear reactor physics. The averaged number of emitted neutrons per one fission reaction is known as the ν value. This value is significantly dependent on a fissioning nuclide and the energy of the incident neutron.

The kinetic energy of neutrons newly generated by fission reactions is much higher than that of thermal neutrons, and the kinetic energy distribution of fission neutrons is dependent on a fissioning nuclide. This energy distribution is known as the fission spectrum and is denoted to as $\chi(E)$. In the case of the fission reaction of ^{235}U induced by thermal neutrons, the averaged energy of fission neutrons is approximately 2 MeV.

One fission reaction of ^{235}U induced by a thermal neutron emits energy of approximately 200 MeV, which is equivalent to 200 MeV $\times$ 1.602 $\times$ 10^{-13} J MeV^{-1} = 3.2 $\times$ 10^{-11} J. If 1 g of ^{235}U, which corresponds to 6.02 $\times$ $10^{23}/235$ = 2.56 $\times$ 10^{21} atoms, fissions, the emitted energy is 8.21 $\times$ 10^{10} J, and this energy is equivalent to approximately 22,800 kWh or 1 MWd.

The ^{238}U, whose concentration in the natural uranium is around 99.3%, does not fission with thermal neutrons, but it can fission with neutrons with high energy above 1 MeV. In addition to the fission reactions of ^{235}U, those of ^{238}U by high energy neutrons occur in nuclear reactors.

In heavier nuclides than uranium, fission reactions can spontaneously occur without any neutron interactions by the tunnel effect known in the quantum mechanics. This fission reaction is known as the spontaneous fission. For example, 1 g of californium-252 can spontaneously fission approximately 6.2 $\times$ 10^{11} times and can emit approximately 2.3 $\times$ 10^{12} neutrons per 1 s.

2.2 Reaction Cross Sections

In this section, a physical quantity, the cross section, is introduced in order to quantitatively discuss the interactions between neutrons and nuclei in nuclear reactors. This quantity can be interpreted as an occurrence probability of specific types of interactions between neutrons and nuclei. Values of cross sections are dependent on the target nuclide in the interaction and the kinetic energy of the incident neutron.

In the following, two physical quantities, the microscopic cross sections and the macroscopic cross sections, are introduced. The former is defined for nuclides and the latter is for the media or materials comprising nuclear reactors.

2.2.1 *Microscopic Cross Sections*

Let us consider a uniform streaming of neutrons moving to a certain direction with a certain energy, and let the number of neutrons passing through a unit-area surface per unit second I (# $\mathrm{cm}^{-2}\ \mathrm{s}^{-1}$). As shown in Fig. 2.1, these neutrons pass through a thin plate-like target vertically. The thickness of this target is assumed sufficiently small; it is guaranteed that the number of neutrons is not decreased through passing the target material. This target is composed of unique nuclide and number density of this nuclide in the target is N (# cm^{-3}). Now we assume that the number of times of specific reaction (i.e., capture reaction) is R per unit volume and per unit time.

If the thickness of this target is dx, the number of reactions occurring in the target having unit surface area and thickness of dx per unit second, Rdx, should be proportional to I, N, and dx, and then the following equation can be defined with the coefficient σ:

$$Rdx = \sigma INdx, \text{ and thus } R = \sigma IN \tag{2.1}$$

This equation can be rewritten as

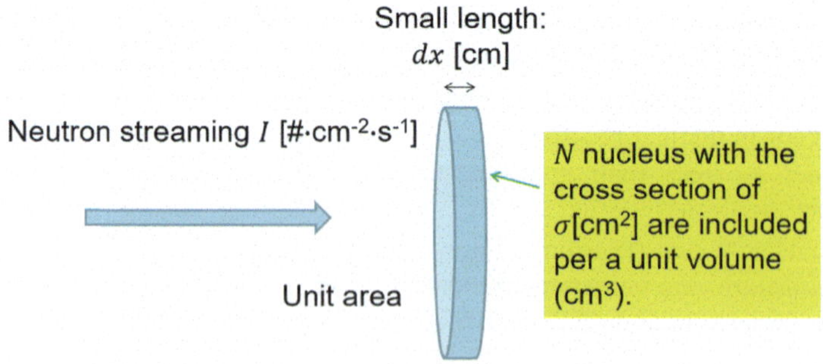

Fig. 2.1 Neutrons passing through a thin plate-like target

$$\sigma = \frac{R/N}{I}. \tag{2.2}$$

This suggests that the coefficient σ corresponds to the expected number of reactions per one nucleus when one neutron passes through the material per unit second and per unit area. This coefficient σ is known as the microscopic cross section. It has a dimension of area, and the units such as cm^2 are used. The other unit of barn (denoted to as b) is also used and $1\ b = 10^{-24}\ cm^2$.

The origin of the terminology "cross section" can be understood if Eq. (2.2) is rewritten as

$$N\sigma = \frac{R}{I}. \tag{2.3}$$

If all neutrons passing through the material contribute to the reactions, the number of reactions should be equal to the number of the entering neutrons. The right-hand side of Eq. (2.3) is a ratio of the number of neutrons causing reactions to the total number of entering neutrons. In other words, an effective area of nuclei in the material per unit surface area is $R/I = N\sigma$. The target having unit surface area and unit thickness should contain N nuclei. Now we consider a very thick material, so overlap of nuclei against the neutron streaming can be neglected. Thus σ can be interpreted as an effective area of one nucleus for the concerned reaction. It can be also said that each nucleus has the effective area of σ (cm^2), and a point-like neutron causes the interaction when passing through this area.

2.2.2 *Macroscopic Cross Sections*

Let us consider that the mono-direction neutron beam is injected into a thick slab as shown in Fig. 2.2 and the number of neutrons entering the slab surface per unit second and per unit area is I (# $cm^{-2}\ s^{-1}$). Now we are concerning a change in the number of streaming neutrons within small width of dx at the depth of x (cm) from the surface. It is assumed that only neutron capture reaction can occur in this medium.

If the number of streaming neutrons at x is denoted to as $I(x)$, the number of neutron capture reactions occurring from x to $x + dx$ is $\sigma I(x)Ndx$ as discussed in Sect. 2.2.1, and this should be equal to a decrease in the number of streaming neutrons, $-dI(x)$. Thus, the following equation should be preserved:

$$-dI(x) = \sigma I(x)Ndx. \tag{2.4}$$

From the above equation, the following differential equation can be derived:

$$\frac{dI(x)}{dx} = -\sigma NI(x). \tag{2.5}$$

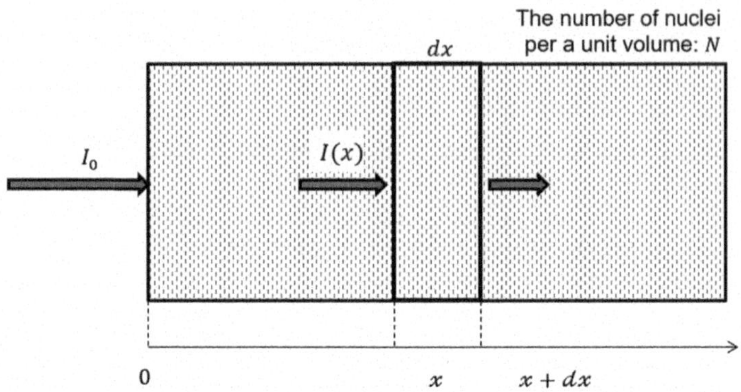

Fig. 2.2 Neutron streaming through a thick material plate

The solution to this differential equation can be written as

$$I(x) = I_0 \exp(-\sigma N x), \tag{2.6}$$

where I_0 is the number of streaming neutrons at the material surface, $x = 0$. A product Nx in this equation is known as the macroscopic cross section and is denoted to as Σ. Its dimension is an inverse of length, and a unit of cm^{-1} is generally used. The number of streaming neutrons in this material can be represented with Σ as

$$I(x) = I_0 \exp(-\Sigma dx). \tag{2.7}$$

Equation (2.5) can be rewritten as follows by using the macroscopic cross section Σ:

$$-\frac{dI(x)}{I(x)} = \Sigma dx. \tag{2.8}$$

This equation suggests that Σ corresponds to a decrease rate in the number of streaming neutrons per unit length.

The macroscopic cross section Σ is defined as σN as described above, but this is for medium consisting of the unique nuclide. If medium composed of several different nuclides is concerned, the macroscopic cross section is defined as follows. If we let the microscopic cross section and the number density of the nuclide i in the medium σ_i and N_i, respectively, the macroscopic cross section of this medium is defined as $\Sigma = \sum_i N_i \sigma_i$.

Since the macroscopic cross sections are defined dependent on the reaction type, a subscript for the reaction type is generally added to Σ. For example, the macroscopic cross sections for the absorption reaction and the fission reaction are denoted to as Σ_a and Σ_f, respectively. This rule on the subscript is commonlty adopted to the microscopic cross sections.

2.2.3 Energy Dependence of Cross Sections

As previously described, reaction cross section values are dependent on the target nuclide with which a neutron interacts, the reaction type, and the energy of the incident neutron. As an example, the cross sections for several reactions of ^{235}U and ^{238}U are shown in Fig. 2.3. These cross section data are taken from the evaluation in the evaluated nuclear data library JENDL-5 [1]. Cross sections of other heavy nuclides behave similarly with those of ^{235}U and ^{238}U.

In a low energy range below 1 eV, the cross sections are almost proportional to $1/v(\propto \sqrt{1/E})$. Above 1 eV, the cross sections are significantly dependent on the incident neutron energy. Each peak observed in the cross section curves is known as the resonance, and this cross section behavior can be explained by a fact that the neutron-nucleus reaction can occur very easily when the incident neutron energy matches with the excitation level of the compound nucleus (the target nucleus taking in neutron). In the resonance energy range, peaks of the resonances become low and the width of the resonances become narrow with the increase of the incident neutron energy, and finally neighboring resonances overlap with each other. Above approximately 1 keV, each resonance cannot be resolved, and finally the cross section curves look smooth. In a high energy range above 1 MeV, the fission cross sections increase and the capture cross sections decrease with the increase in the incident neutron energy.

Behavior of the fission cross sections of ^{235}U and ^{238}U in a high energy range is very different from each other. That of ^{238}U shows very low values in a low energy range, but it drastically increases with high incident neutron energy above 1 MeV and finally it becomes comparable with those of ^{235}U.

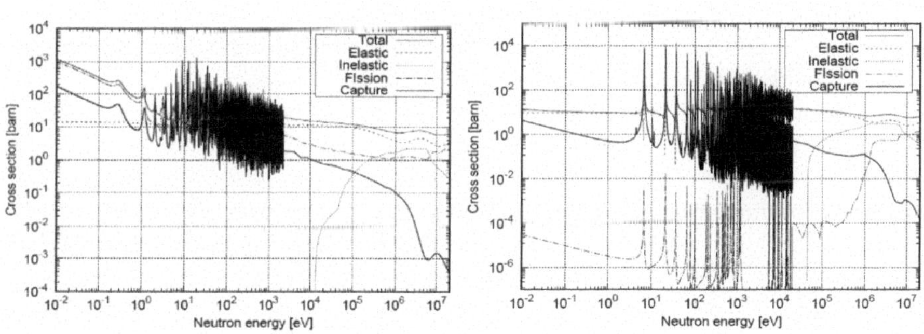

Fig. 2.3 Reaction cross sections of ^{235}U (left) and ^{238}U (right)

2.3 Fission Chain Reaction: Neutron Multiplication Factor and Six-Factor Formula

The nuclear fission reaction yields two or three new neutrons in most cases, and thus if these neutrons can induce subsequent fission reactions, fission chain reactions can be realized. Nuclear reactors are systems utilizing and realizing this fission chain reaction.

In a critical state, the number of fission reactions occurring in a nuclear reactor is almost constant with time without any neutron supply from outside of the system. The number of fission reactions decreases with time in a subcritical state, and it increases in a supercritical state. Let us consider how these critical, subcritical, and supercritical states are attained in a certain condition.

In order to understand this, let us consider an entire life of one neutron in a nuclear reactor, which is born by a fission reaction. Here thermal neutron reactors are concerned. Neutrons generated by fission reactions have high energy of approximately 2 MeV on average. They experience multiple collisions with nuclei, which are mainly light nuclides such as hydrogen and oxygen, and lose their kinetic energy. This process is known as the neutron slowing-down. Finally, they become thermal neutrons having low energy of approximately 0.025 eV, are absorbed by nuclei in nuclear fuels, and cause the fission reactions.

Let us preview a physical process that fast neutrons generated by the fission reactions by the thermal neutrons cause the next fission reaction in Fig. 2.4. A small fraction of fast neutrons generated by fission reactions cause new fission reactions and yield new fast neutrons; the number of fast neutrons is slightly increased due to this. Then, these fast neutrons become thermal neutrons and cause next fission reactions when they satisfy the following conditions:

- During the slowing-down process of fast neutrons to become thermal neutrons, they should not escape from a system and continue to stay.
- During the slowing-down process, neutrons should not be captured by heavy nucleus; they should escape from the resonance captures of heavy nuclides such as ^{238}U.
- Thermalized neutrons should not escape from a system and continue to stay.
- Thermalized neutrons should not be absorbed by nucleus in non-fuel media such as moderator and structural materials and should be absorbed by fuel media.
- When thermalized neutrons are absorbed by nucleus of fuel media, fission reactions should selectively occur.

Neutrons satisfying the above conditions can cause the next fission reactions. Per one fission reaction, ν fast neutrons are generated on average, and they experience the above process. This physical process repeatedly occurs, and fission chain reactions continue.

A terminology, generation, is defined as a time duration of the above one-cycle process: from the birth of a fast neutron by a fission reaction to a next fission reaction which thermalized neutron causes. A ratio of the number of neutrons existing in

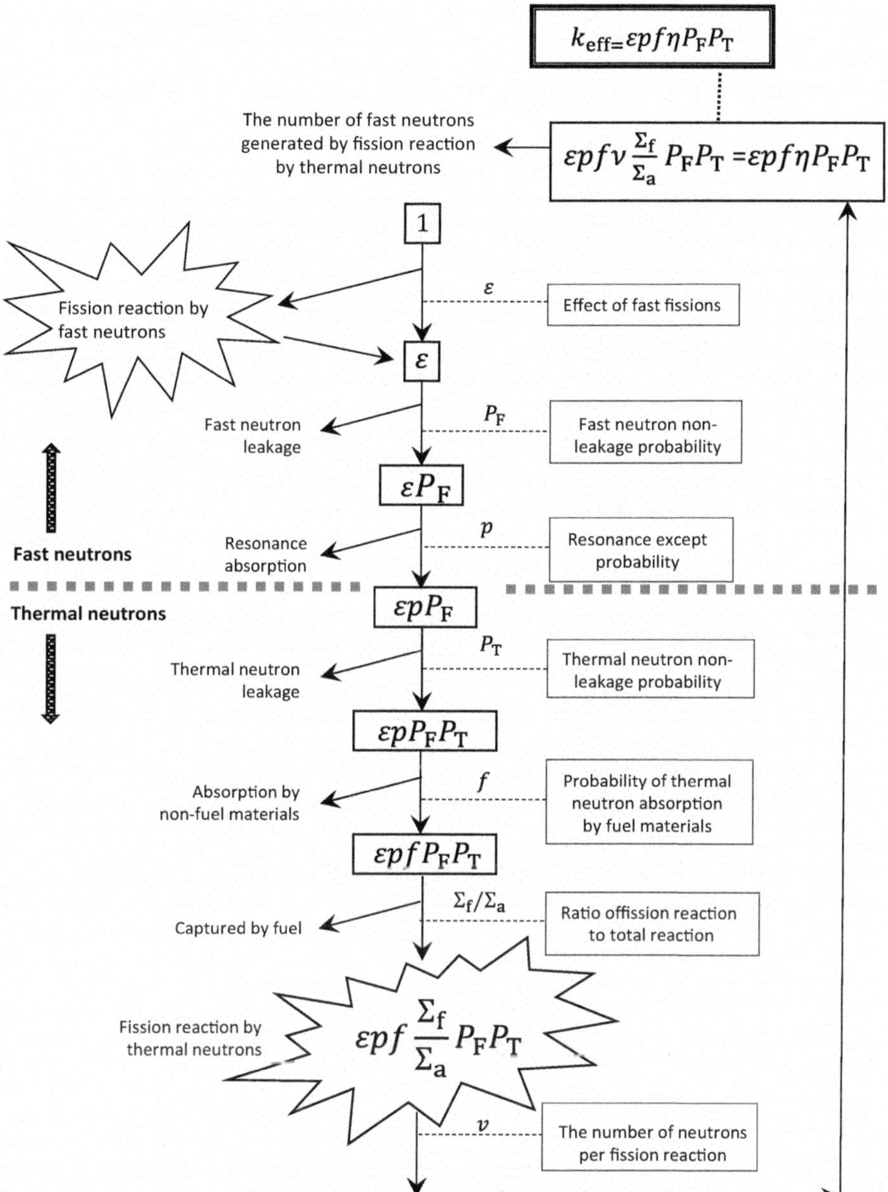

Fig. 2.4 One-cycle physical process occurring during fission chain reaction

a reactor at a certain generation to that at the preceding generation is known as a neutron multiplication factor k:

$$k = \frac{\text{The number of neturons in a certain generation}}{\text{The number of neutrons in a preceding generation}}.$$

Let us assume that a certain number of neutrons is instantaneously introduced to a system somehow and consider behavior of these neutrons with time according to the value of neutron multiplication factor k.

- In the case of $k = 1$, the number of neutrons is constant at every generation, so the number of neutrons initially introduced to a system is unchanged with generation; after the introduction of neutrons, fission chain reactions sustain without any further introduction of neutrons. This state is known as a critical state.
- In the case of $k < 1$, the number of neutrons decreases with the generation, and thus the fission chain reaction is not sustainable if any neutrons are not introduced. This state is known as a subcritical state.
- In the case of $k > 1$, the number of neutrons exponentially increases with generation. This state is known as a supercritical state (Fig. 2.5).

As described above, behavior of the number of neutrons in a system is significantly dependent on whether k is equal to/less than/larger than unity.

Next, let us consider several physical phenomena which one neutron experiences during one generation and attempt to describe each of them in detail.

(1) Fast neutron fission factor ε

Neutrons generated by fission reactions are fast neutrons having 2 MeV on average. Some of these fast neutrons are absorbed by nuclei in nuclear fuels without slowing-down and induce the next fission reactions especially by ^{238}U which can fission only with fast neutrons. In thermal neutron reactors using natural or low-enriched uranium, this ^{238}U fission should be taken into account. Since more than one neutron are generated by fission reactions with fast neutrons, the number of neutrons at

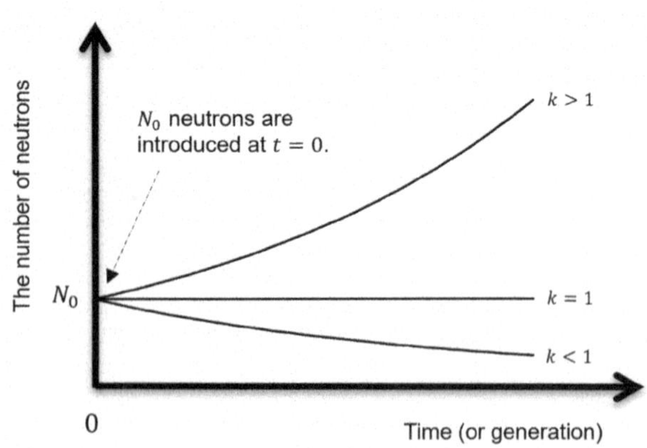

Fig. 2.5 Time behavior of the number of neutrons dependent on the neutron multiplication factor

the initial stage of one generation is effectively increased, and this effect can be considered by using the following fast neutron fission factor ε:

$$\varepsilon = \frac{\text{The number of fission neturons by fast and thermal neutrons}}{\text{The number of fission neutrons by thermal neutrons}}.$$

(2) Fast neutron non-leakage probability P_F

Some of fast neutrons generated by fission reactions leak from a system during the slowing-down and do not contribute to the next fission reactions. This effect can be considered by the fast neutron non-leakage probability P_F.

(3) Resonance escape probability p

Fast neutrons generated by fission reactors lose their kinetic energy through the elastic scattering with nucleus in the moderator, mainly hydrogen, and become thermal neutrons. During this slowing-down process, some neutrons are lost from a system by the resonance absorptions of heavy nuclei, mainly ^{238}U. The probability that fast neutrons escape from the resonance absorptions and are successfully thermalized is defined as the resonance escape probability p.

(4) Thermal neutron non-leakage probability P_T

Some of thermalized neutrons leak from a system and do not contribute to the next fission reactions. This is accounted for by the thermal neutron non-leakage probability P_T.

(5) Thermal neutron utilization factor f

Lives of thermal neutrons which do not leak from a system are terminated by absorptions of nuclei in fuel materials or non-fuel materials such as moderator and structural materials. The probability that thermal neutrons are absorbed by the fuel material is defined as the thermal neutron utilization factor f.

(6) Neutron reproduction factor η

When thermal neutrons are absorbed by the fuel materials, sometimes neutrons are just captured and fission reactions are not induced. In the nuclear reactor physics, the neutron reproduction factor η is defined as the expected number of newly-generated neutrons per one neutron absorption by the fuel material: a product of a probability of fission reaction occurrence with thermal neutron absorptions by fuel material and the expected number of neutrons generated by one fission reaction ν.

Using the above-defined six factors, we can find that one neutron generated by a fission reaction induced by a thermal neutron yields $\varepsilon p f \eta P_F P_T$ neutrons at the next generation. Thus, neutron multiplication factor k can be represented as

$$k = \varepsilon p f \eta P_F P_T. \tag{2.9}$$

In this equation, the neutron multiplication factor is expressed as a product of the six factors, and thus this equation is known as the six-factor formula.

If a system is infinite (or is expanded over an infinite space), there should not be leakage of fast and thermal neutrons and both of P_F and P_T are unity. In such a situation, the neutron multiplication factor is represented as $\varepsilon pf\eta$, and this multiplication factor defined for an infinite system is known as the infinite multiplication factor k_∞:

$$k_\infty = \varepsilon pf\eta. \tag{2.10}$$

This equation is known as the four-factor formula for the infinite multiplication factor.

On the other hand, the neutron multiplication factor k for a system in which neutron leakage from a system exists is known as the effective multiplication factor k_{eff}. On these two factors, k_{eff} and k_∞, the following relation should hold:

$$k_{eff} = \varepsilon pf\eta P_F P_T = k_\infty P_F P_T. \tag{2.11}$$

Based on the Fermi age theory and the neutron diffusion theory described in Sect. 2.5.2, the neutron non-leakage probabilities P_F and P_T can be represented as

$$P_F = e^{-B^2\tau_T}, \tag{2.12}$$

$$P_T = \frac{\Sigma_a}{\Sigma_a + DB^2} = \frac{1}{1 + L_T^2 B^2}, \tag{2.13}$$

where τ_T is the Fermi age of thermal neutrons, whose dimension is a square of length, and L_T^2 is the diffusion area, a square of the diffusion length L_T. The parameter B^2 is the buckling which corresponds to the degree of the distortion in neutron flux spatial distribution in a system, and is significantly relevant to neutron leakage from a system. This will be discussed in Sect. 2.5.

Based on the above discussion, we can obtain the following equation:

$$k_{eff} = \frac{k_\infty e^{-B^2\tau_T}}{1 + L_T^2 B^2}. \tag{2.14}$$

2.4 Reaction Rate, Neutron Flux, and Neutron Balance Equation

In this section, terminologies, the reaction rate and the neutron flux, are introduced and briefly described. The reaction rate is to quantify the number of reactions between neutrons and nuclei, and the neutron flux is a helpful quantity to define the reaction rate. In addition, the balance equation on the reaction rate in a system is derived as the neutron diffusion equation.

2.4.1 Reaction Rate and Neutron Flux

Let us consider a neutron "group" moving to a certain direction with a velocity v (cm s^{-1}) in a system, and let the number of neutrons per unit volume, which is also known as the neutron density, n (cm^{-3}). Here let us consider the number of reactions between these neutrons and nuclei existing in this system.

If we consider a unit-area surface vertical to the velocity of the neutron group, vn neutrons should pass through the surface per unit time. A probability of the reaction occurrence in a volume from the surface to the depth of small width dx is given as Σdx where Σ is a macroscopic cross section. Thus the number of reactions between neutrons and nucleus in this volume having unit area surface and thickness of dx per unit time can be represented as $vn\Sigma dx$. This is the reaction rate. If we consider a volume having unit surface and unit thickness (or unit volume), the number of reactions between neutrons and nucleus in this unit volume is $vn\Sigma$. A product vn (cm^{-2} s^{-1}) is known as the neutron flux, and is denoted to as

$$\phi = vn(\mathrm{cm}^{-2}\,\mathrm{s}^{-1}). \tag{2.15}$$

Using the neutron flux, the reaction rate can be represented as $\Sigma\phi$. This discussion suggests that the neutron flux can be also interpreted as a mathematical quantity to quantify the reaction rate.

In nuclear reactors, neutrons should move to various directions and thus the dependence of the neutrons on the moving direction should be taken into account. At a certain spatial position, the neutron flux directing to small solid angle $d\Omega$ along to a direction Ω is known as the angular neutron flux. The unit of this quantity is cm^{-2} s^{-1} sr^{-1}, and is denoted to as ψ in the following. The neutron flux described in the preceding sections is formally known as the scalar neutron flux, and the scalar neutron flux ϕ is defined from the angular neutron flux ψ as

$$\phi(\mathbf{r}, E) = \int_{4\pi} \psi(\mathbf{r}, E, \Omega) d\Omega. \tag{2.16}$$

Let us remember that we have discussed the neutron streaming in Sect. 2.2. Actually this corresponds to the angular neutron flux. When we consider streaming of the neutron group, it is convenient if we can define a unique quantity summarizing the angular neutron fluxes depending on the directions. If neutrons move randomly to various directions and there is no specific direction to which most neutrons in the group move at a certain spatial position, it can be said that there is no bias in the streaming. The streaming of the neutron group can be defined as a vector quantity, the neutron current, and is denoted to as $\boldsymbol{J}$ in the following. Rigorously, the neutron current is defined from the angular neutron flux as

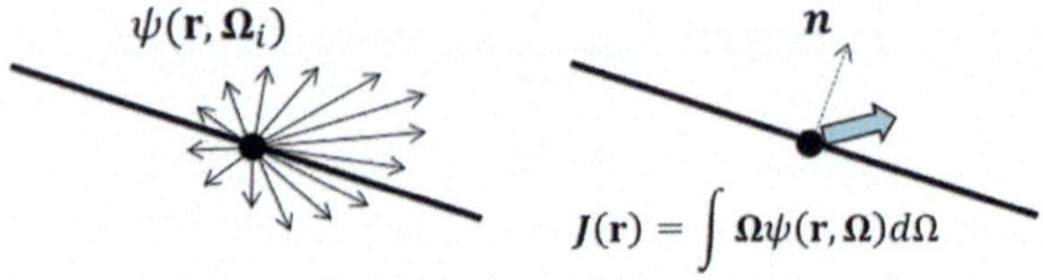

Fig. 2.6 Angular neutron flux and neutron current

$$J(\mathbf{r}, E) = \int_{4\pi} \Omega\psi(\mathbf{r}, E, \Omega)d\Omega. \tag{2.17}$$

Let us consider a surface with unit area perpendicular to the neutron current direction at a spatial position **r**. Some of neutrons crossing this surface should move to the same direction as the current ($J \cdot \Omega > 0$), and the others should move to the opposite direction ($J \cdot \Omega < 0$). If the numbers of neutrons passing through this surface per unit time are denoted to as J_+ for the former and as J_- for the latter, the net number of neutrons passing through this surface directing to the vector J can be represented as $(J_+ - J_-)$. Actually, this corresponds to the length (or size) of the neutron current vector J, $|J|$, whereas the rigorous derivation is omitted here. The length of the neutron current is the net number of neutrons crossing a unit-area surface perpendicular to the vector per a unit time.

Next, let us consider a unit-area surface independent on the direction of the neutron current vector at a certain spatial location, and the normal vector to this surface is represented as n as shown in Fig. 2.6. The net number of neutrons crossing this surface per unit time is given as $|J|$ if the directions of n and J are exactly same with each other. If there is a difference in the directions between n and J, the net number of the crossing neutrons is represented as the n-direction component of J, $(n \cdot J)$. When a three-dimensional xyz-coordinate system is concerned, the neutron current vector can be represented as $\left(J_x, J_y, J_z\right)$. In this case, the net number of neutrons crossing the unit-area yz-plane per unit time is J_x, for example.

2.4.2 Balance Equation of Neutron Density and Neutron Diffusion Equation

In this section, a balance of the neutron density in a small volume within a nuclear reactor is discussed. Qualitatively speaking, a rate in the neutron density change, dn/dt, is a difference between the generation rate and the loss rate by the leakage and absorptions. When the generation rate and the loss rate balance with each other, the neutron density is expected unchanged with time.

Firstly, let us consider the neutron loss rate by leakage. In the preceding section, the net number of neutrons passing through a unit-area surface can be quantified by the neutron current vector J. When we consider a small volume in a space, the net number of neutrons leaking from this volume is quantified as the surface integral of the inner product $\boldsymbol{n} \cdot \boldsymbol{J}$ as follows:

$$\int_A \boldsymbol{n} \cdot \boldsymbol{J} dS, \tag{2.18}$$

where A is a total surface area of this volume. The adoption of the divergent theorem transforms the surface integral into the volume integral as

$$\int_A \boldsymbol{n} \cdot \boldsymbol{J} dS = \int_V \mathrm{div} \boldsymbol{J} dV, \tag{2.19}$$

where V is a volume concerned. Furthermore, if Fick's law is introduced between the neutron current and the neutron flux, the neutron current vector can be represented with the neutron flux as

$$\boldsymbol{J} = -D\,\mathrm{grad}\phi, \tag{2.20}$$

where D is a diffusion coefficient.

This equation suggests that the magnitude of the streaming of the neutron group is proportional to the gradient of the neutron flux spatial distribution and the diffusion coefficient. By substituting Eq. (2.20) into Eq. (2.19), the following equation can be derived:

$$\int_V \mathrm{div}(-D\mathrm{grad}\phi) dV = -\int_V D\mathrm{div}(\mathrm{grad}\phi) dV = -\int_V D\nabla^2\phi dV. \tag{2.21}$$

This is the number of neutrons lost through the leakage from the surface of the concerned volume. If this quantity is negative, the neutron group inflows toward inside of this volume.

The neutron generation rate in this small volume is represented as $\int_V SdV$ when S neutrons are generated per unit volume and per unit time. The loss rate by the neutron absorptions is similarly represented as $\int_V \Sigma_a \phi dV$.

Now we can derive the neutron balance equation based on the above discussion. The rate of the change in the number of neutrons is represented as

$$(\text{Change rate}) = (\text{Generation}) - (\text{Loss by absorption}) - (\text{Loss by leakage}), \tag{2.22}$$

and after each term is replaced by the mathematical expression, the following equation can be derived:

$$\int_V \frac{\partial n}{\partial t} dV = \int_V S dV - \int_V \Sigma_a \phi dV - \left(- \int_V D\nabla^2 \phi dV \right). \tag{2.23}$$

Now we are discussing an arbitrary small volume in a system, and thus the following equation should be preserved at every spatial position in this system:

$$\frac{1}{v} \frac{\partial \phi}{\partial t} = D\nabla^2 \phi - \Sigma_a \phi + S. \tag{2.24}$$

Note that $\phi = vn$ is used. This equation is known as the neutron diffusion equation. If neutrons are generated only by fission reactions, the generation term S can be replaced by $\nu \Sigma_f \phi$ since the number of occurring fission reactions per unit time and per unit volume is $\Sigma_f \phi$ and the averaged number of neutrons generated by one fission reaction is ν. Then, Eq. (2.24) is rewritten as.

$$\frac{1}{v} \frac{\partial \phi}{\partial t} = D\nabla^2 \phi - \Sigma_a \phi + \nu \Sigma_f \phi. \tag{2.25}$$

If a reactor is in a critical state, the time derivative term in the left-hand side of Eq. (2.25) is zero, and thus

$$-D\nabla^2 \phi + \Sigma_a \phi = \nu \Sigma_f \phi. \tag{2.26}$$

This equation is fundamental and important for criticality calculations in the following sections.

In the above derivation of the neutron diffusion equation, all neutrons are assumed to have a unique energy and spatial dependence of the parameters such as D and Σ_a are ignored for simplicity. In actual, neutrons should have various energy and the neutron flux is dependent on both the energy and the spatial position. Parameters such as diffusion coefficients and reaction cross sections should also be dependent on the neutron energy and the position.

Furthermore, the validity of the neutron diffusion approximation is sometimes lost. In such cases, the neutron transport equation considering the energy, spatial position, and moving direction should be treated.

2.5 Buckling and Neutron Non-leakage Probability

2.5.1 Physical Meaning of Buckling

This section provides a detailed explanation on the buckling B^2 which has been introduced in Sect. 2.3 to define the neutron non-leakage probabilities in the six-factor formula.

If the energy dependence of the neutron fluxes and the reaction cross sections is ignored (or the one-energy group approximation is adopted) and the diffusion theory is adopted, the neutron flux in a homogeneous thermal critical reactor satisfies the following equation:

$$-D\nabla^2\phi + \Sigma_a\phi = \nu\Sigma_f\phi. \tag{2.27}$$

This equation is rewritten as

$$\nabla^2\phi + B^2\phi = 0, \tag{2.28}$$

where

$$B^2 = \frac{\nu\Sigma_f - \Sigma_a}{D}. \tag{2.29}$$

The parameter B^2 defined in Eq. (2.29) is known as the buckling. Since this buckling is defined from the parameters D, Σ_a, and $\nu\Sigma_f$ which are relevant to the material comprising the nuclear reactor, this is also known as the material buckling. Since the numerator in Eq. (2.29) is a difference of the fission production cross section from the neutron absorption cross section, the material buckling can be interpreted as the material performance-relevant parameter on the fission chain reactions. When $\nu\Sigma_f = \Sigma_a$, the material buckling is zero and the neutron diffusion equation becomes $-D\nabla^2\phi = 0$. This means that if a nuclear reactor using the fuel whose material buckling is zero reaches a critical state, the neutron leakage from this reactor should be zero (or the size of this reactor should be infinite).

In the following, a rectangular parallelpiped reactor with which our experiment is carried out, is discussed.

Let the extrapolated length of each side a, b, and c in a parallelpiped reactor. Equation (2.28) for this system is written as

$$\frac{\partial^2\phi}{\partial x^2} + \frac{\partial^2\phi}{\partial y^2} + \frac{\partial^2\phi}{\partial z^2} + B^2\phi = 0. \tag{2.30}$$

If uniform boundary conditions are assumed on each surface of this reactor, the neutron flux ϕ can be represented as $\phi(x, y, z) = X(x)Y(y)Z(z)$ using three independent components. By substituting this to Eq. (2.30), the following equation can be derived:

$$\frac{1}{X}\frac{d^2X}{dx^2} + \frac{1}{Y}\frac{d^2Y}{dy^2} + \frac{1}{Z}\frac{d^2Z}{dz^2} + B^2 = 0. \tag{2.31}$$

The first three terms in the left-hand side of Eq. (2.31) are dependent only on x, y, and z, respectively, and the sum of them is constant, $-B^2$. This suggests that each of these three terms should be constant. Thus,

$$\frac{1}{X}\frac{d^2X}{dx^2} = -\alpha^2, \frac{1}{Y}\frac{d^2Y}{dx^2} = -\beta^2, \frac{1}{Z}\frac{d^2Z}{dx^2} = -\gamma^2, \tag{2.32}$$

where $\alpha^2 + \beta^2 + \gamma^2 = B^2$.

Now we have three independent equations for $X(x)$, $Y(y)$, and $Z(z)$, and these are solved by adopting the surface boundary conditions that the neutron flux at the extrapolated position is zero. In the case of $X(x)$, the general solution to the corresponding differential Eq. (2.32) can be obtained as $\cos \alpha x$ and $\sin \alpha x$. If we set the origin of the coordinate system at a reactor central position, the surface boundary condition gives $\alpha a/2 = (2k+1)\pi/2$ for $\cos \alpha x$, and $\alpha a/2 = k\pi$ for $\sin \alpha x$ where k is an arbitral natural number. Thus, the following equations are obtained for $X(x)$:

$$X_l(x) = \begin{cases} A_l \cos \frac{l\pi x}{a}, l = 1, 3, 5, \ldots \\ A_l \sin \frac{l\pi x}{a}, l = 2, 4, 6, \ldots \end{cases}, \tag{2.33}$$

where A_l is a constant. On the functions $Y(y)$ and $Z(z)$, the same procedure leads

$$Y_m(x) = \begin{cases} A'_m \cos \frac{m\pi x}{b}, m = 1, 3, 5, \ldots \\ A'_m \sin \frac{m\pi x}{b}, m = 2, 4, 6, \ldots \end{cases}, \tag{2.34}$$

$$Z_n(x) = \begin{cases} A''_n \cos \frac{n\pi x}{c}, n = 1, 3, 5, \ldots \\ A''_n \sin \frac{n\pi x}{c}, n = 2, 4, 6, \ldots \end{cases}. \tag{2.35}$$

Now we know that

$$\alpha = \left(\frac{l\pi}{a}\right), \beta = \left(\frac{m\pi}{b}\right), \gamma = \left(\frac{n\pi}{c}\right), \tag{2.36}$$

and the corresponding B^2 is obtained as

$$B_{lmn}^2 = \left(\frac{l\pi}{a}\right)^2 + \left(\frac{m\pi}{b}\right)^2 + \left(\frac{n\pi}{c}\right)^2. \tag{2.37}$$

The fundamental mode is defined as the smallest one of B^2 in Eq. (2.37) where $l = m = n = 1$, and other ones are known as the higher-order modes. The fundamental mode solution is written as

$$\phi_{111}(x, y, z) \propto \cos \frac{\pi x}{a} \cos \frac{\pi y}{b} \cos \frac{\pi z}{c}. \tag{2.38}$$

To better understand a difference between the fundamental and higher-order mode components, let us consider a supercritical reactor without an external neutron source and this reactor is made in a critical state by the control rod manipulation. Before this operation, the neutron flux level should increase with time with a specific spatial distribution. If the control rod is inserted to this supercritical reactor, a neutron flux spatial distribution should initially include the fundamental mode and several higher-order modes, but the higher-order modes decrease with time, and finally the neutron flux spatial distribution should be converged to that of the fundamental mode.[1]

Now we are discussing the neutron flux spatial distribution in a critical state: the fundamental mode. Thus, the buckling B^2 should satisfy the specific relation with the reactor size-relevant parameters a, b, and c as follows:

$$B^2 = \left(\frac{\pi}{a}\right)^2 + \left(\frac{\pi}{b}\right)^2 + \left(\frac{\pi}{c}\right)^2. \tag{2.39}$$

The buckling defined from the reactor size-relevant parameters is known as the geometric buckling. As Eq. (2.39) clearly shows, the geometric buckling of a parallelpiped reactor should decrease with the increase in the reactor size. This is general in other reactors having different geometric specification. Since neutron leakage should be decreased when the reactor size increases, the geometric buckling should decrease when the neutron leakage from a reactor becomes small.

In the above discussion, the material-relevant parameters such as D, Σ_a, and $\nu\Sigma_f$ are assumed known for a critical reactor. Thus the material buckling can be obtained from these parameters with Eq. (2.29), and the size of this critical reactor should satisfy Eq. (2.39). Oppositely, if a critical reactor with a certain size is given in advance, the geometric buckling of this reactor can be obtained according to Eq. (2.39). If this reactor is critical, the material buckling of this fuel should satisfy Eq. (2.29).

As described above, we have known that the material buckling corresponds to a parameter to quantify the fuel material performance about the fission chain reaction and can take both positive and negative values. Also we have known that the geometric buckling corresponds to a parameter to quantify the degree of the neutron leakage from a reactor and can take only non-negative values. If a reactor is in a critical state, the material buckling of the loaded fuel and the geometric buckling of this reactor should be identical to each other. If the geometric buckling is larger than the material buckling, it means that the neutron leakage is superior to the fuel performance, and thus the reactor should be in a subcritical state, and vice versa. The geometric buckling is non-negative, so nuclear reactors loading the fuel material with negative material buckling can never reach a critical state.

[1] Rigorously speaking, the higher-order modes described here are different from those defined in this section. More precisely, the higher-order modes defined in this section are not for a critical reactor, but for a reactor in a certain-level supercritical state.

2.5.2 Derivation of Neutron Non-leakage Probability in Six-Factor Formula

In this section, Eqs. (2.12) and (2.13) for the neutron non-leakage probabilities in the six-factor formula, P_{F} and P_{T}, are derived.

Let us consider the thermal neutron leakage firstly. When the thermal neutron flux ϕ_{T} satisfies an equation similar to the one-group diffusion equation, the total amount of the thermal neutron leakage from a system is represented as $-D\int_V \nabla^2\phi_{\mathrm{T}}dV$. This can be rewritten with the geometric buckling as

$$-D\int_V \nabla^2\phi_{\mathrm{T}}dV = DB^2\int_V \phi_{\mathrm{T}}dV. \tag{2.40}$$

The total number of the thermal neutrons lost from a system by the absorptions is equal to $\Sigma_{\mathrm{a}}\int_V \phi_{\mathrm{T}}dV$. Since the thermal neutrons in a nuclear reactor should be lost through the leakage or absorption, the thermal neutron non-leakage probability P_{T} should be represented as

$$P_{\mathrm{T}} = 1 - \frac{DB^2\int_V \phi_{\mathrm{T}}dV}{\Sigma_{\mathrm{a}}\int_V \phi_{\mathrm{T}}dV + DB\int_V \phi_{\mathrm{T}}dV} = \frac{\Sigma_{\mathrm{a}}\int_V \phi_{\mathrm{T}}dV}{\Sigma_{\mathrm{a}}\int_V \phi_{\mathrm{T}}dV + DB\int_V \phi_{\mathrm{T}}dV}. \tag{2.41}$$

The integral terms in the numerator and the denominator are dropped and the thermal neutron diffusion area $L_{\mathrm{T}}^2 = D/\Sigma_{\mathrm{a}}$ are introduced, and then

$$P_{\mathrm{T}} = \frac{\Sigma_{\mathrm{a}}}{\Sigma_{\mathrm{a}} + DB^2} = \frac{1}{1 + L_{\mathrm{T}}^2B^2}. \tag{2.42}$$

Next let us consider the leakage of neutrons during the slowing-down process from the fission energy to the thermal energy. The total number of fast neutrons generated in a nuclear reactor per unit time is represented as

$$\begin{aligned} \varepsilon\int_V \nu\Sigma_{\mathrm{f}}\phi_{\mathrm{T}}dV &= \varepsilon\Sigma_{\mathrm{a}}\frac{\Sigma_{\mathrm{a,fuel}}}{\Sigma_{\mathrm{a}}}\frac{\nu\Sigma_{\mathrm{f}}}{\Sigma_{\mathrm{a,fuel}}}\int_V \phi_{\mathrm{T}}dV = \varepsilon\Sigma_{\mathrm{a}}f\eta\int_V \phi_{\mathrm{T}}dV \\ &= \frac{k_\infty}{p}\Sigma_{\mathrm{a}}\int_V \phi_{\mathrm{T}}dV. \end{aligned} \tag{2.43}$$

Based on the Fermi age theory, the number of the thermalized neutrons per unit second in an idealized system without any resonance absorption can be represented with the neutron slowing-down density q_{T} as $\int_V q_{\mathrm{T}}dV$. By using a relation between the slowing-down density and the thermal neutron flux [2],

$$q_T(\boldsymbol{r}) = \frac{k_\infty}{p}\Sigma_{\mathrm{a}}e^{-B^2\tau_{\mathrm{T}}}\phi_{\mathrm{T}}(\boldsymbol{r}), \tag{2.44}$$

the number of the thermalized neutrons in a reactor per unit time is written as

$$\frac{k_\infty}{p}\Sigma_{\mathrm{a}} e^{-B^2\tau_{\mathrm{T}}}\int_V \phi_{\mathrm{T}} dV. \tag{2.45}$$

The probability that fast neutrons are thermalized without any leakage from a system P_{F} should be defined as a ratio of the number of thermalized neutrons to the number of the fast original neutrons generated by fission neutrons, and thus

$$P_{\mathrm{F}} = \frac{\frac{k_\infty}{p}\Sigma_{\mathrm{a}} e^{-B^2\tau_{\mathrm{T}}}\int_V \phi_{\mathrm{T}} dV}{\frac{k_\infty}{p}\Sigma_{\mathrm{a}}\int_V \phi_{\mathrm{T}} dV} = e^{-B^2\tau_{\mathrm{T}}}. \tag{2.46}$$

2.6 Nuclear Reactor Kinetics

2.6.1 *Reactivity*

The reactivity is a parameter to quantify how far a nuclear reactor is from a critical state, and is generally denoted to as ρ. The formal definition of the reactivity is

$$\rho = \frac{k_{\mathrm{eff}} - 1}{k_{\mathrm{eff}}}. \tag{2.47}$$

The reactivity ρ is dimensionless, but a unit of $\Delta k/k$ is practically used. Other units of $\%\Delta k/k$ for $10^{-2}\Delta k/k$ and pcm for $10^{-5}\Delta k/k$ are also used. A reactivity normalized by the effective delayed neutron fraction β_{eff}, which is explained later, is also a reactivity with a unit of the dollar, and one cent is 1/100 of one dollar.

The reactivity corresponds to a state of a nuclear reactor. A difference between two different reactor states is known as the reactivity worth, and its unit is $\Delta k/kk'$. Practically the terminology reactivity is used for both the reactivity and the reactivity worth.

2.6.2 *Prompt and Delayed Neutrons*

Some of fission fragments and their daughter nuclei generated by fission reactions can emit one neutron through the β decay. Neutrons generated through this mechanism are known as the delayed neutrons while neutrons generated simultaneously by the fission reactions are known as the prompt neutrons. Nuclides which can emit delayed neutrons through the β decay, such as ^{87}Br, are known as the delayed

Table 2.1 Delayed neutron data of ^{235}U fission by thermal neutrons [3]

Precursor group	Decay constant λ_i (sec^{-1})	Relative abundance ($a_i = \beta_i/\beta$)
1	0.0124 ± 0.0003	0.033 ± 0.003
2	0.0305 ± 0.0010	0.219 ± 0.009
3	0.111 ± 0.004	0.196 ± 0.022
4	0.301 ± 0.011	0.395 ± 0.011
5	1.14 ± 0.15	0.115 ± 0.009
6	3.01 ± 0.29	0.042 ± 0.008

neutron precursors. Neutron emissions from the precursors occur according to their decay half-lives. The number of the precursor nuclides is larger than 300, and these precursors are categorized into several groups according to their half-lives when the nuclear kinetics is discussed. The number of the precursor groups is generally six. As an example, the six-group data of precursors generated by ^{235}U fission reactions with thermal neutrons are shown in Table 2.1, where the decay constants and the relative abundances which are ratios of precursor group-wise precursor yields to the total yield are provided.

An expected ratio of the number of delayed neutrons to that of total neutrons in a fission reaction is known as the delayed neutron fraction, and is generally denoted to as β. Energy of the delayed neutrons emitted from the precursors is generally lower than that of the prompt neutrons, so the non-leakage probability of the delayed neutrons during the slowing-down is relatively high. Thus, the possibility of the delayed neutrons causing the next fission reaction is higher than that of the prompt neutrons. By considering this effect, the effective delayed neutron fraction is introduced and is denoted to as β_{eff}. Values of β_{eff} are dependent on the reactor core configuration, fuel composition, neutron energy spectra, etc.

2.6.3 Prompt Neutron Lifetime and Neutron Generation Time

Prompt neutrons generated by fission reactions are absorbed by nucleus in reactor materials or leak from a reactor, and terminate their lives. Averaged time duration from the prompt neutron generation to its end of life is known as the prompt neutron lifetime and is denoted to as l. Values of the prompt neutron lifetime in several different types of nuclear reactors are presented in Table 2.2. Those in thermal neutron reactors range from 5×10^{-5} to 10^{-3} s, and those are dependent on the moderator materials. The prompt neutron lifetime of fast reactors is shorter by orders of two to five than those of thermal reactors.

In addition to the prompt neutron lifetime, the neutron generation time Λ is also used. This parameter can be interpreted as an average time duration required for the generation of one neutron and $l = \Lambda \times k_{\text{eff}}$. When l and Λ are identical with each

Table 2.2 Prompt neutron lifetime of various reactor types

Reactor type	Moderator	Reactor name	Prompt neutron lifetime (second)
Thermal reactors	Graphite	Calder hall	10^{-3}
	Heavy-water (CANDU)	Douglas point-1 (Canada)	6.5×10^{-4}
	Heavy-water	JRR-2 (JAEA)	2.1×10^{-4}
	Light-water	KUR	6.7×10^{-5}
	Light-water	KUCA C45G0	5.5×10^{-5}
Fast reactors		Enrico Fermi	1.4×10^{-7}
		EBR-2	8×10^{-8}
		YAYOI	2.9×10^{-8}

other, one neutron is newly generated during the same time duration for which one neutron ends its life, and this should correspond to $k_{\text{eff}} = 1.0$.

2.6.4 Derivation of Point Kinetics Equation

Let us consider a kinetics behavior of an infinite-sized nuclear reactor with the six-delayed neutron precursor group model.

Let the number of the ith-group precursors per unit volume, that is the precursor density, $C_i(t)$ and its decay constant λ_i. Thus, the number of the precursors which decay per unit time and per unit volume is represented as $\lambda_i C_i(t)$. The number of the generated ith-group precursors can be represented as $\beta_{\text{eff},i} \nu \Sigma_{\text{f}} \phi(t)$, so the following balance equation on $C_i(t)$ can be derived for each precursor group:

$$\frac{dC_i(t)}{dt} = -\lambda_i C_i(t) + \beta_{\text{eff},i} \nu \Sigma_{\text{f}} \phi(t), \ (i = 1, 2, \ldots, 6). \tag{2.48}$$

The neutron balance equation with considering the delayed neutrons can be represented as follows since an infinite-sized reactor is now concerned:

$$\frac{dn(t)}{dt} = -\Sigma_{\text{a}} \phi(t) + (1 - \beta_{\text{eff}}) \nu \Sigma_{\text{f}} \phi(t) + \sum_{i=1}^{6} \lambda_i C_i(t), \tag{2.49}$$

where $n(t)$ is a neutron density per unit volume. The left-hand side of Eq. (2.49) is a temporal change in the neutron density; the first term of the right-hand side is the number of neutrons disappearing by absorption; the second term is the number of generated prompt neutrons; the third term is the number of generated delayed neutrons by the decay of the precursors.

By using the relation between $n(t)$ and the scalar neutron flux $\phi(t)$, $\phi(t) = vn(t)$ where v is the neutron speed, Eq. (2.49) is transformed to

$$\begin{aligned}\frac{dn(t)}{dt} &= v\{-\Sigma_a + (1-\beta_{\mathrm{eff}})\nu\Sigma_f\}n(t) + \sum_{i=1}^{6}\lambda_i C_i(t) \\ &= v\Sigma_a\left\{(1-\beta_{\mathrm{eff}})\frac{\nu\Sigma_f}{\Sigma_a} - 1\right\}n(t) + \sum_{i=1}^{6}\lambda_i C_i(t).\end{aligned} \tag{2.50}$$

By using the prompt neutron lifetime ℓ, the neutron multiplication factor k, and the reactivity ρ, which are defined as

$$l = \frac{1}{v\Sigma_a}, \tag{2.51}$$

$$k = \frac{\nu\Sigma_f}{\Sigma_a}, \tag{2.52}$$

$$\rho = \frac{k-1}{k}, \tag{2.53}$$

Equation (2.50) is further transformed into

$$\frac{dn(t)}{dt} = \frac{(1-\beta_{\mathrm{eff}})k - 1}{l}n(t) + \sum_{i=1}^{6}\lambda_i C_i(t) = k\frac{\rho - \beta_{\mathrm{eff}}}{l}n(t) + \sum_{i=1}^{6}\lambda_i C_i(t). \tag{2.54}$$

Similarly, Eq. (2.48) on the delayed neutron precursors becomes

$$\frac{dC_i(t)}{dt} = -\lambda_i C_i(t) + \beta_{\mathrm{eff},i} v\Sigma_a \frac{\nu\Sigma_f}{\Sigma_a}n(t) = -\lambda_i C_i(t) + \beta_{\mathrm{eff},i}\frac{k}{l}n(t). \tag{2.55}$$

The simultaneous equations on $n(t)$ and $C_i(t)$, Eqs. (2.54) and (2.55), are known as the point kinetics equation. In the above discussions, an infinite-sized reactor has been concerned. If a finite-sized reactor is concerned, the similar equation can be derived by assuming that the neutron flux spatial distribution is unchanged during the transient and the magnitude (or level) of the neutron flux changes with time: the spatially- and time-dependent neutron flux $\Phi(\boldsymbol{r}, t)$ can be represented as $\Phi(\boldsymbol{r}, t) = \phi(t)f(\boldsymbol{r})$ where $\phi(t)$ and $f(\boldsymbol{r})$ are the time-dependent and spatially-dependent components of the neutron flux, respectively.

By replacing the prompt neutron lifetime l by the neutron generation time Λ, Eqs. (2.54) and (2.55) are transformed into the following equations, respectively, which are also generally used as the point kinetics equation:

$$\frac{dn}{dt} = \frac{\rho - \beta_{\text{eff}}}{\Lambda} n + \sum_{i}^{6} \lambda_i C_i, \tag{2.56}$$

$$\frac{dC_i}{dt} = \frac{\beta_{i,\text{eff}}}{\Lambda} n - \lambda_i C_i. \tag{2.57}$$

Let us discuss each term of the point kinetics equation above. Equation (2.56) is to present the change in the neutron density in a nuclear reactor; the term in the left-hand side is for a change rate in the neutron density; the first term in the right-hand side is for a change due to the loss by the absorption/leakage and the generation of the prompt neutrons; the second term in the right-hand side is for a change due to the delayed neutron generation by the precursor decay. Equation (2.57) is to present the change in the precursor densities; the term in the left-hand side is a change rate of the ith group precursor density; the first term in the right-hand side is for a change due to generation by the fission reaction) the second term in the right-hand side is for a change due to the loss by the precursor decay.

At the beginning of the present chapter, we have never considered the delayed neutrons. This is because we discussed the behavior of the number of neutrons in a stationary state in which no change is expected in the number of neutrons, and the neutron generation rate and the loss rate balance with each other. On the other hand, when we need to discuss the neutron kinetics problem in which the number of neutrons changes with time, consideration of the delayed neutrons and their precursors is mandatory.

By using Eq. (2.54), the importance of the delayed neutrons in reactor kinetics problems can be easily demonstrated. If we assume that all neutrons generated by fission reactions are prompt neutrons, the following equation can be derived since $\beta_{\text{eff}} = 0$ in Eq. (2.54):

$$\frac{dn(t)}{dt} = \frac{k-1}{l} n(t). \tag{2.58}$$

This equation can be easily solved with the initial condition of $n(0) = n_0$ as

$$n(t) = n_0 \exp\left(\frac{k-1}{l} t\right). \tag{2.59}$$

Here, let us consider a case where the neutron multiplication factor k changes by only $+0.01\%$ from a critical state ($k = 1$). This is almost equivalent to a perturbation where a control rod is slightly moved. By assuming $l \approx 5 \times 10^{-5}$ (s) as KUCA, we can obtain

$$\frac{k-1}{l} t = \frac{1.0001 - 1}{5 \times 10^{-5}} t = 2t, \tag{2.60}$$

and $n(t) = n_0 \exp(2t)$. This means that the neutron flux increases over 7 times during only one second, and it is easily known that such reactors cannot be controlled by humankinds. Actually, 0.7% of fission neutrons are delayed neutrons which are generated with the time delay from fission reactions, so the control of the system becomes possible if k is increased by about 0.01%.

In the case of $0 < \rho < \beta_{\text{eff}}$, the first term of the right-hand side in Eq. (2.56) is negative, so the change in $n(t)$ should be negative by the contribution of this term. However, the second term is positive and this positively contributes to the change in $n(t)$. The contribution of the second term is mild, so change in $n(t)$ is also mild when $n(t)$ increases with time. On the other hand, in the case of $\rho > \beta_{\text{eff}}$, the change in $n(t)$ is quite rapid since the contribution of the first term relevant to the prompt neutrons is dominant. The situation of $\rho = \beta_{\text{eff}}$ is called a prompt critical, and that of $\rho > \beta_{\text{eff}}$ is called a prompt supercritical. The latter should be avoided in nuclear reactor operations. Sometimes, the situation of $\rho > \beta_{\text{eff}}$ is also called a prompt supercritical.

Generally, the situation of $\rho = 0$ is called a critical state, but it is also called as a delayed critical to distinguish it from the prompt critical.

In the criticality accident occurring at the JCO Tokai branch in September, 1999, through pouring uranyl-nitrate ($UO_2\,(NO_3)_2$) solution to a tank, the reactivity of this neutron multiplicative system reached approximately 3 dollars ($\rho = 3\beta_{\text{eff}}$), and the rapid fission chain reaction was initiated. A word "criticality" in the terminology, the criticality accident, is rigorously a prompt critical state. Generally, the terminology "criticality accident" is used when reactivity of systems reaches larger values than the delayed critical ($\rho \geq 0$) in facilities in which a subcritical state ($\rho < 0$) should be maintained at non-reactor facilities such as nuclear fuel fabrication facilities.

2.6.5 Solution to Point Kinetics Equation

When the reactivity ρ, the effective delayed neutron fractions $\beta_{\text{eff},i}$ and the prompt neutron lifetime l are constant with time, $n(t)$ can be represented as

$$n(t) = \sum_{j=1}^{7} A_j e^{\omega_j t}. \tag{2.61}$$

By substituting Eq. (2.61) into the kinetics Eqs. (2.56) and (2.57), the following equation which ω_j should satisfy can be derived:

$$\rho = \frac{\omega l}{1 + \omega l} + \frac{\omega}{1 + \omega l} \sum_{i=1}^{6} \frac{\beta_{\text{eff},i}}{\omega + \lambda_i}. \tag{2.62}$$

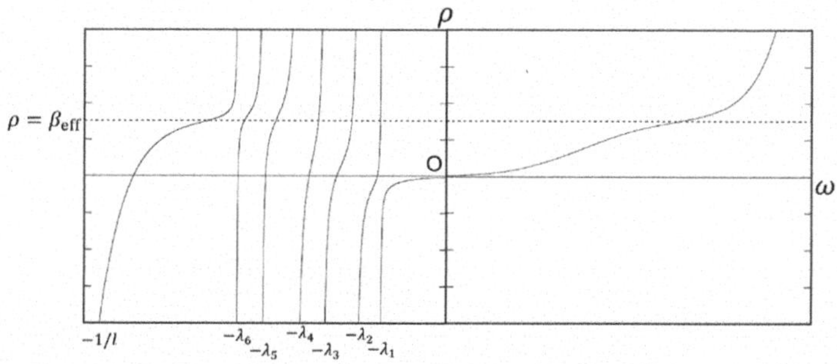

Fig. 2.7 Relation between ρ and ω in the reactivity equation

This is known as the reactivity equation. Parameters ω_j in Eq. (2.61) correspond to roots of the reactivity equation at the given reactivity ρ. Figure 2.7 shows an example of the relation between ρ and ω in Eq. (2.62). The x-axis of this figure is log-scale like: in the range of $\omega < 0$, the parameter in the x-axis corresponds to $-\log(|\omega|)$.

The parameters ω_j, the roots of Eq. (2.62) in the given reactivity $\hat{\rho}$, are presented as crossing points between the curves in this figure and a horizontal line of $\rho = \hat{\rho}$. If the roots are written as $\omega_1, \omega_2, \ldots$ with the descending order, the following observations can be drawn from this figure:

- When a reactivity is larger than zero, ω_1 is positive, and the other ω_j s are negative.
- When a reactivity is zero (or a critical state is assumed), ω_1 is zero and the other ω_j s are negative. The ω_1 component in $n(t)$ becomes $e^{0t} = 1$, and thus $n(t)$ should contain a constant component.
- When a negative reactivity is inserted, all ω_j s are negative.
- When a positive reactivity which is smaller than β_{eff} is inserted, the absolute value of ω_7 becomes extremely large. On the other hand, when a positive reactivity over β_{eff} is inserted, the absolute value of ω_7 becomes small, but that of ω_1 becomes extremely large.

Figure 2.8 is based on the same data as those for Fig. 2.7, but the ranges are different; the reactivity is from 0 to 1 dollar and $\omega > 0$. When the reactivity is 0.8 dollar, ω_1 becomes around 1.0. In this case, the ω_1 component in $n(t)$ increases by e times (2.7 times) during one second, and thus after 10 s, $n(t)$ becomes 22,000 times. This suggests that even though a positive reactivity less than 1 dollar is inserted to a system, the number of neutrons (i.e., the reactor power) should significantly increase during short time duration.

When the contribution of the delayed neutrons is ignored in the kinetics equation for $n(t)$ in Eq. (2.54), the following equation can be derived:

$$\frac{dn(t)}{dt} = \frac{(1 - \beta_{\text{eff}})k - 1}{l} n(t). \tag{2.63}$$

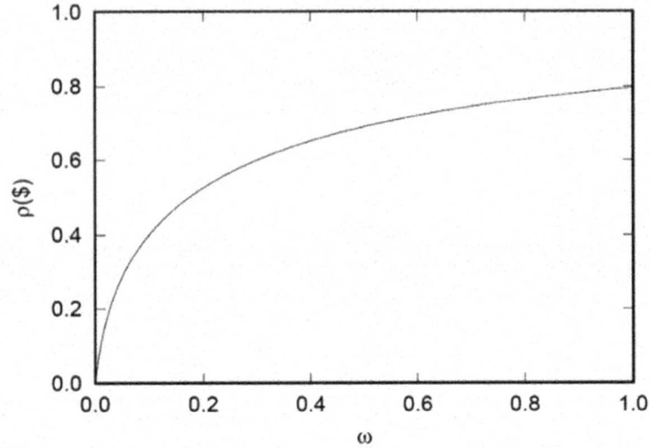

Fig. 2.8 Relation between ρ and ω in the reactivity equation with short range

The general solution to this equation is easily obtained as

$$n(t) = C\exp\left(\frac{(1-\beta_{\mathrm{eff}})k-1}{l}t\right) = C\exp(\hat{w}t), \tag{2.64}$$

where C is an unknown constant.

The parameter $\hat{w}$ in this equation is a good approximation to the absolutely-largest root of the reactivity equation. Figure 2.9 is the same as Fig. 2.7 with different ranges of the reactivity and roots of the reactivity equation, and the curves of "Approx." are obtained from Eq. (2.64). It is well demonstrated that the parameter $\hat{w}$ is a good approximation to ω_7 in $\rho < \beta_{\mathrm{eff}}$ and that to ω_1 in $\rho > \beta_{\mathrm{eff}}$.

The solution to Eqs. (2.56) and (2.57) is given as follows if a reactivity of ρ is inserted to a reactor in a critical state with the number of neutrons of n_0:

$$n(t) = n_0 \sum_{j=1}^{7} \frac{\rho e^{\omega_j t}}{\left\{\rho - \omega_j^2 \sum_{i=1}^{6} \frac{\beta_{i,\mathrm{eff}}}{(\lambda_i+\omega_j)^2}\right\}}. \tag{2.65}$$

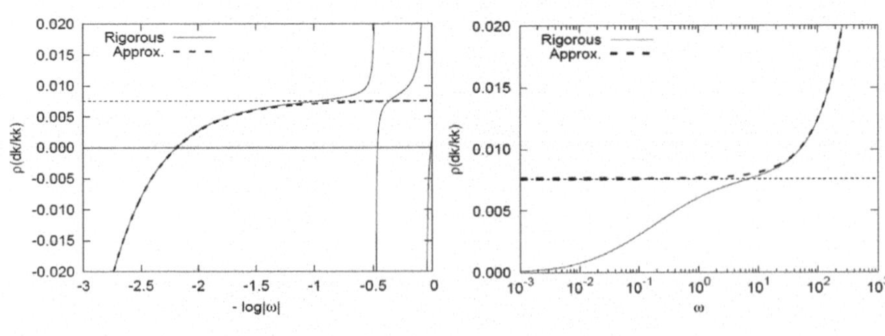

Fig. 2.9 Relation between ρ and $\omega/\hat{w}$ in the reactivity equation

References

1. Iwamoto O, Iwamoto N, Kunieda S, Minato F, Nakayama S, Abe Y et al (2023) Japanese evaluated nuclear data library version 5: JENDL-5. J Nucl Sci Technol 60:1–60. https://doi.org/10.1080/00223131.2022.2141903(OpenAccess)
2. Lamarsh JR (1966) Introduction to nuclear reactor theory. Addison-Wesley Publishing Company, Reading, Massachusetts
3. Keepin GR, Wimett TF, Zeigler RK (1957) Delayed neutrons from fissionable isotopes of uranium, plutonium and thorium, LA-2118, Los Alamos National Laboratory

Chapter 3
Approach-to-Criticality Experiment

Abstract In nuclear reactors, sustainable fission chain reactions by neutrons are realized without any neutron supply from external sources. In other words, nuclear reactors are systems to generate energy and to utilize radiation rays such as neutrons through properly controlling sustainable fission chain reactions. Based on this consideration, the first step to study nuclear reactors would be to know or understand which conditions in geometry, composition, size, etc. are required to make reactors in a critical state, how reactors can reach a critical state, and how reactors behave during approaching to a critical state. For these objectives, the approach-to-criticality experiment, in which a subcritical reactor is made approaching to a critical state, is carried out. This experiment is fundamental and mandatory, and it is generally carried out prior to other nuclear reactor experiments. In the approach-to-criticality experiment at KUCA, fuel plates made of nuclear materials such as uranium are loaded in a reactor core. Neutron flux level in the core is measured at several different subcritical states, number of fuel plates to be added at the next state is determined according to the measurement results obtained at the present state, and finally the reactor reaches a critical state. This experiment is also carried out as preparation for the subsequent experiments such as the control rod calibration experiment. The critical state in which the effective neutron multiplication factor k_{eff} of the core is unity can be clearly defined in theory or numerical simulation. On the other hand, in actual reactors, it is quite difficult to distinguish the critical state from the slightly subcritical or supercritical state when the reactor power is low. If the reactor power is increased, a deviation from the critical state can be clearly observed in the detector signal on the reactor power. One of the purposes of this experiment is to deepen the understanding of the critical state through the actual use of nuclear reactors.

Keywords Neutron multiplication · Inverse multiplication · Inverse count rate · Excess reactivity

C. H. Pyeon et al., *Reactor Laboratory Experiments at Kyoto University Critical Assembly*, https://doi.org/10.1007/978-981-97-8070-9_3

3.1 Principle

3.1.1 Neutron Multiplication in Subcritical Reactor Core

When a nuclear reactor is in a subcritical state without any external neutron sources, the number of neutrons in the core decreases with time and finally becomes zero since fission chain reaction is not sustainable. If neutrons are continuously supplied from external sources, however, this subcritical reactor keeps constant power with time. Here let us consider the multiplication of neutrons supplied from external sources when a reactor is in a subcritical state. For simplicity, the concept of generation for a neutron group is introduced to represent time sequence; the generation is defined here as averaged time duration for which neutrons existing in a reactor are replaced by other neutrons newly generated by fission reactions.

Let us assume that the number of neutrons supplied from external sources per one generation is S. These S source neutrons introduced at the first generation are multiplied by fission reactions, and the number of neutrons at the next generation becomes Sk_{eff}. These Sk_{eff} neutrons are further multiplied and become Sk_{eff}^2 neutrons at the subsequent generation. After the n generations, the number of neutrons becomes Sk_{eff}^n.

Let us consider the total number of neutrons in a subcritical reactor core with external neutron sources at a certain generation when long time passes after the external neutron sources are introduced. Since S neutrons are supplied to the reactor core from the external neutron sources per one generation, there are neutrons originating from the external sources introduced at the current generation, those of one generation before, those of two generations before, etc., as shown in Fig. 3.1. The total number of neutrons at the present generation can be calculated as a sum of them and is represented as

$$S + Sk_{\text{eff}} + Sk_{\text{eff}}^2 + Sk_{\text{eff}}^3 + \cdots = \frac{S}{1 - k_{\text{eff}}}. \tag{3.1}$$

In other words, S neutrons from the external source are multiplied to $\frac{S}{1-k_{\text{eff}}}$ by a system whose neutron multiplication factor is k_{eff} (< 1). The rate of this multiplication,

$$M = \frac{1}{1 - k_{\text{eff}}}, \tag{3.2}$$

is an important parameter in the approach-to-critical experiment. Based on the above description, it becomes possible to know the neutron multiplication factor k_{eff} of the system by measuring how much the source neutron is multiplied, or measuring M, after introducing the external source into a subcritical reactor.

In the above, the equation on the neutron multiplication has been qualitatively derived. Rigorously speaking, the neutron multiplication should be discussed based

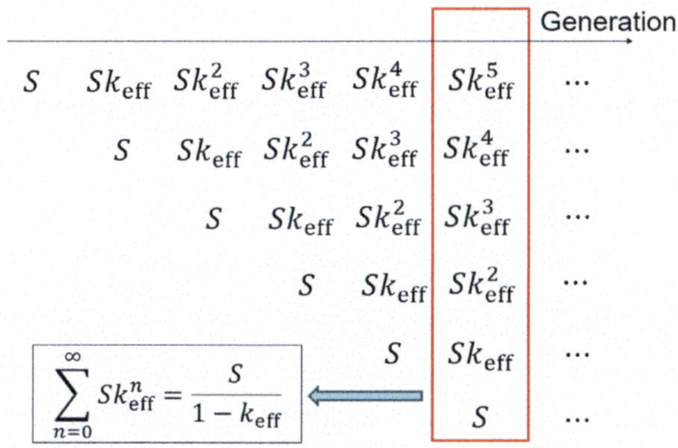

Fig. 3.1 The total number of neutrons at a certain generation

on the theory for bare thermal neutron reactors, but finally the same equation $M = 1/(1-k_{eff})$ is obtained.

3.1.2 Relation Between Inverse Multiplication and Amount of Loaded Fuel

In the approach-to-criticality experiment, external neutron source is introduced to a subcritical system in which k_{eff} is sufficiently smaller than unity at first. Then k_{eff} is gradually approached to unity observing the change in M defined in Eq. (3.2), and finally a state in which $k_{eff} = 1$ (or $M \to \infty$) is realized. While there are various ways to make k_{eff} close to unity (or make a subcritical reactor close to a critical state) and these are dependent on a reactor type, a procedure to increase the amount of nuclear fuels loaded to a reactor core is adopted in this experiment.[1]

The definition of M tells us that it is not easy to handle M itself since this quantity becomes infinity in a critical state, and thus the inverse of M is treated instead in the approach-to-criticality experiment. It is known as the inverse multiplication and is defined as

$$\frac{1}{M} = 1 - k_{eff}. \tag{3.3}$$

Based on this definition, the inverse multiplication becomes unity when there is no neutron multiplication in a system ($k_{eff} = 0$) and zero when a system is in a critical state ($k_{eff} = 1$).

Generally, k_{eff} increases with the increase of the amount of the loaded nuclear fuels; if the inverse multiplication $1/M$ is plotted against the amount of loaded

[1] IN commercial power reactors, the amount of the loaded nuclear fuels is determined through its design, so the approach to criticality is attained by adjusting the boron concentration in coolant or changing control rod insertion positions after the predetermined amount of nuclear fuels are loaded.

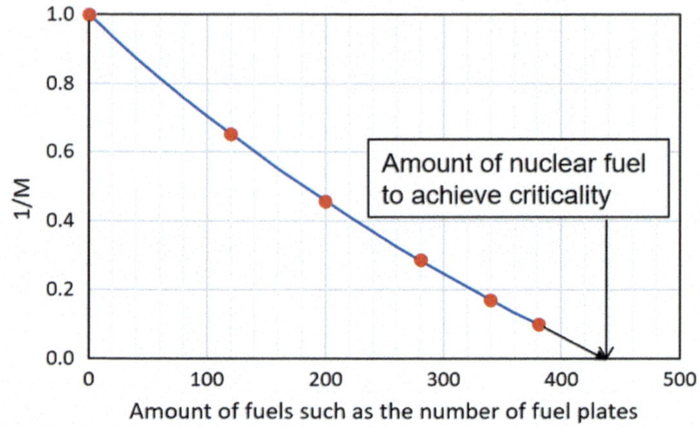

Fig. 3.2 Inverse multiplication against the amount of loaded nuclear fuels

nuclear fuels, it should give a curve which crosses the y-axis at $y = 1$ and crosses the x-axis at the amount of the loaded fuels with which a critical state can be achieved as shown in Fig. 3.2. As a conclusion, estimation of the amount of loaded fuels required to achieve the critical state is possible with the following procedure:

- k_{eff} is increased by increasing the amount of the loaded nuclear fuels,
- $1/M$ is measured,
- A curve for $1/M$ is plotted with the amount of the loaded fuels as the x-axis and $1/M$ as the y-axis, and,
- An x value where the $1/M$ curve crosses the x-axis is estimated by the extrapolation.

3.1.3 Measurement of Inverse Multiplication Ratio

When we rigorously follow the procedure described in the preceding section to obtain the amount of the loaded nuclear fuels required to achieve the critical state based on the quantity $1/M$, we have to start the experiment with a state in which no nuclear fuels are loaded in the core (i.e., $k_{\mathrm{eff}} = 0$).

On the other hand, in the experiment at KUCA, certain amount of nuclear fuels are loaded at the initial state. This initial core is called the reference core. In this case, the procedure mentioned above cannot be used as is. In the following, we discuss another quantity rather than the inverse multiplication.

Let the effective neutron multiplication factors at the reference core and at the other core with the loaded state i, $k_{\mathrm{eff},0}$ and $k_{\mathrm{eff},i}$, respectively. Under the respective states, neutron fluxes ϕ_0 and ϕ_i are roughly represented as consequences of the multiplication of the source neutrons S as

$$\phi_0 \propto \frac{S}{1 - k_{\text{eff},0}}, \tag{3.4}$$

$$\phi_i \propto \frac{S}{1 - k_{\text{eff},i}}. \tag{3.5}$$

When the amount of the loaded nuclear fuels is increased from the reference core 0 to the core with the state i, the ratio of neutron flux through this change ϕ_i/ϕ_o is represented as

$$\frac{\phi_i}{\phi_0} = \frac{\frac{S}{1-k_{\text{eff},i}}}{\frac{S}{1-k_{\text{eff},0}}} = \frac{1 - k_{\text{eff},0}}{1 - k_{\text{eff},i}}. \tag{3.6}$$

The reference core 0 is common for the arbitrary fuel-loaded state i, so $1 - k_{\text{eff},0}$ is constant, and thus the above equation is written as

$$\frac{\phi_i}{\phi_0} \propto \frac{1}{1 - k_{\text{eff},i}}, \tag{3.7}$$

and its inverse is written as

$$\frac{\phi_0}{\phi_i} \propto 1 - k_{\text{eff},i}. \tag{3.8}$$

These equations take the same form of M or $1/M$; If the neutron flux levels are measured at the initial state and the fuel loaded state i, and the ratio of them ϕ_0/ϕ_i can be obtained, the same procedure for the inverse multiplication $1/M$ can be adopted without any modifications.

In the actual experiments, neutron flux levels are measured by neutron detectors such as fission chambers, BF_3 detectors, ^3He detectors, etc. These neutron detectors measure the neutron flux level as reaction rates between neutrons and specific materials, and the count rates are obtained depending on the efficiency inherent to each detector. In the approach-to-criticality experiment at KUCA, count rates measured by the neutron detector A_0 and A_i are assumed proportional to the neutron flux level as

$$\frac{A_0}{A_i} \propto \frac{\phi_0}{\phi_i} \propto 1 - k_{\text{eff},i}. \tag{3.9}$$

The inverse of the ratio of the neutron count rates A_0/A_i is known as the inverse count rate.

As a conclusion, the approach-to-criticality experiment is carried out with the following procedure to achieve the critical state:

- To prepare a reactor core with a loaded state i, nuclear fuel plates are added to a reactor core with a preceding state and k_{eff} is increased.

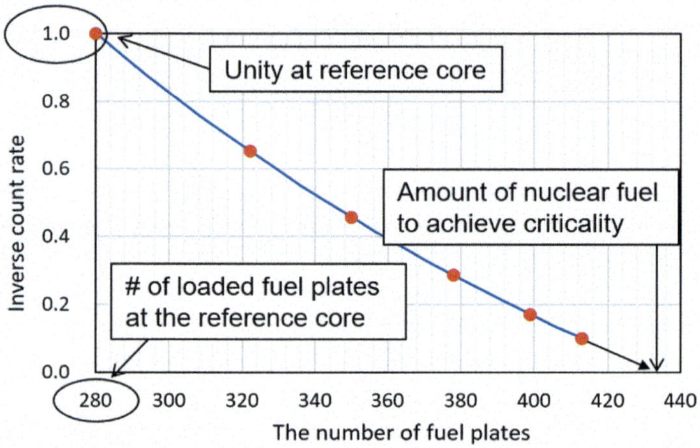

Fig. 3.3 Inverse count rate curve

- Count rates of the neutron detectors A_i are obtained through the measurement.
- As shown in Fig. 3.3, the curve of the inverse count rate is plotted on the figure with the number of loaded fuel plates as the x-axis and the inverse count rates as the y-axis.
- Through obtaining the cross point of the inverse count rate curve and the x-axis by the extrapolation, the number of fuel plates required to achieve the criticality is estimated and determined.

3.2 Experimental Methods

3.2.1 Measurement Procedure

General cautions during the experiments are described in the following:

- The loading of nuclear fuel plates into fuel frames is carried out in the fuel room. This operation should be done carefully by considering the cautions to handle with the nuclear fuel and radioactive materials.
- In the operation to change the reactor power such as insertion and withdrawal of an external source, insertion and withdrawal of control and safety rods, flooding and draining of the water in the reactor core tank, etc., it is important to take care of the changes in the indications of neutron detection systems and to well grasp the current core state.
- It is strongly recommended to confirm the relation in the spatial positions among the neutron detectors used in the measurement, the reactor core, and the external neutron source from documents provided by instructers and the core monitoring

system in the control room. The fuel reloading operation would be also a beneficial occasion for this.

In the following description, nuclear fuel loading states in a core are represented by the index i. The reference core is denoted as $i = 0$, and this index is increased as $i = 1, 2, \cdots$ with the increase of the amount of the loaded fuels.

(1) Preparation of the reference core

To construct the reference core, the proper amount of nuclear fuels is loaded in a core, and the neutron detectors are also properly located. This preparation is generally carried out prior to the experiment, so when the experiment begins, the amount of the nuclear fuels loaded in the core and the positions of the neutron detectors and the external source should be checked in advance.

(2) Reactor start-up

The external neutron source is inserted into the core, the three safety rods (S4 to S6) are withdrawn, and the reactor core tank is flooded with the water whose level is 1450 mm.

(3) Change of the control rod insertion pattern and neutron count measurement

In this experiment, neutron count rates or detector power level are measured at three different control rod insertion patterns as shown in Table 3.1 at every fuel loading state. The control rod insertion pattern is represented by the index j.

At each fuel loading state, measurement data of the neutron count rate are obtained by each of the neutron detectors, and the measured count rates at the fuel loading state i with the control rod insertion pattern j are represented as A_{ij}. The following should be noted:

- Neutron count measurement is carried out using the timer and scaler when the fission chamber and the auxiliary detection systems are employed. The measurement time is determined according to the count rates, and from the viewpoint of the statistical uncertainties, it is recommended to set the timer to obtain the neutron counts larger than 1000. When the reactor core is approaching a critical state, the neutron count rates become large, and thus the measurement time should be changed to save the overall time for the experiment.

Table 3.1 Three patterns of control rod positions in the approach-to-criticality experiment

Pattern index	Control rod		
	C1	C2	C3
$j = 1$	Full insertion (0 mm)	Full insertion (0 mm)	Full insertion (0 mm)
$j = 2$	Two: full withdrawal (650 mm), one: full insertion (0 mm)*		
$j = 3$	Full withdrawal (650 mm)	Full withdrawal (650 mm)	Full withdrawal (650 mm)

*Which control rods are fully inserted/withdrawn are dependent on the reactor core in the experiment

- The measurement error in the count rate at the reference core $i = 0$, A_{0j}, affects the values of the inverse multiplication at all states since A_{0j} is used to calculate the inverse multiplication.

Detailed procedures of the control rod insertion pattern modification and the neutron count measurement are described in the following:

- $j = 1$: When a nuclear reactor is started up and the core tank is filled with water, all three control rods are fully inserted. This reactor state is denoted to as $j = 1$. After confirming the reactor power becomes stable, neutron count measurement is started.
- $j = 2$: Preassigned two control rods are withdrawn, and the state of $j = 2$ in which two control rods are fully withdrawn and the remaining control rod is fully inserted, is attained. Observe the change in the reactor power carefully during the control rod withdrawal. As well as the case at $j = 1$, neutron count measurement is carried out after confirming that the reactor power becomes stable.
- $j = 3$: One fully-inserted control rod is withdrawn and the state of $j = 3$ is attained. Observe the change in the reactor power during the control rod withdrawal. If the reactor is still in a subcritical state, neutron count measurement is started after confirming that the reactor power is stable. At the final step in the approach-to-criticality experiment, generally, a reactor becomes critical during the withdrawal of the remaining control rod before attaining the state of $j = 3$. How to confirm that the reactor becomes critical and how to obtain the measurement data will be explained in the procedure (8).

(4) Preparation of the inverse count rate curve

After the neutron count measurement, the inverse count rate curve with the number of the loaded fuel plates as the x-axis and the inverse count rate A_{0j}/A_{ij} as the y-axis is prepared for each neutron detector and each control rod insertion pattern $j = 1$ to 3. Three different curves dependent on the control rod insertion pattern should be obtained.

(5) Temporal stop of the reactor operation

When the neutron count measurement is terminated and it is confirmed that the obtained measurement data are reasonable, the reactor core operation is temporarily terminated by the withdrawal of the external source, insertion of three control rods, and drain of the water in the reactor tank. Positions of the three safety rods are kept at the top position. After this procedure, discharging and reloading of the nuclear fuels become possible.

Also, during the water drain, it is possible to observe the change in the reactor power by the linear power monitor and to obtain the measurement data for the exercise raised in the Sect. 3.3.2. Water drain operation can be done by manipulating the flooding/draining operation lever. The initial water level is 1450 mm. Spatial relation between the water level and the fuel frame can be found in Fig. 1.8 in Chap. 1.

This measurement is carried out with the following way:

- On the log data of the linear power monitor, water levels with a 10 mm difference from the initial value of 1450 mm is marked such as 1450, 1440 mm, ...
- When the water level reaches 1250 mm, the measurement is finished.
- After the experiment, a copy of the log data will be distributed. Using this, prepare a graph of the power level as the y-axis with the water level of the x-axis.

(6) Determination of the number of added fuel plates

To proceed the next experimental step, the number of added fuel plates should be determined. In this experiment, the number of the loaded fuel plates at the first several steps is predetermined. At the steps where the number of added fuel plates is not predetermined, the number of the fuel plates required to attain the criticality is estimated through the extrapolation of the two latest values of the inverse count rate; a cross point of the extrapolated line and the x-axis should be the number of the fuel plates for the criticality achievement. This should be done by each of experimental groups using the measurement results obtained with their preassigned detectors.

After estimating the number of fuel plates for the criticality achievement, share the information on the results with the trend in the inverse count rate curve with each of the experimental groups. Then, have a discussion among all the participants to determine the number of the added nuclear fuel plates. In this discussion, report their own results with their interpretations based on the spatial relation among the reactor core, the detectors, and the external source, and the direction along which the fuel plates are added. The number of actually added fuel plates should be smaller than half of the difference in the number of fuel plates between at the current state and at the predicted critical state until the strong confidence on the estimation is derived. At the final step of the approach-to-criticality experiment, after confirming that the reactivity increase by the next fuel reloading can be controlled (or compensated) by one control rod, which is withdrawn at the control rod patten change from step $j = 2$ to 3, the fuels are reloaded to make the reactor a slight supercritical state, and a critical state is attained by adjusting the control rod position.

Various trends such as a convex-upward shape, a convex-downward shape, and a straight line can be observed in the inverse count rate curve dependent on the concerned detector. Due to this, the predicted number of the fuel plates for the criticality can be extremely different from the true value as shown in Fig. 3.4 if the reactor is far from a critical state. When determining the number of added fuel plates by all participants, all the information provided by all the experimental groups should be considered.

(7) Loading of nuclear fuels to a reactor core

Loaded fuel frames containing fuel plates are discharged from a core, the fuel plates with the determined number are added to the fuel frame in the fuel room, and the fuel frames are reloaded to the core.

The above procedure from (2) to (7) is repeatedly carried out.

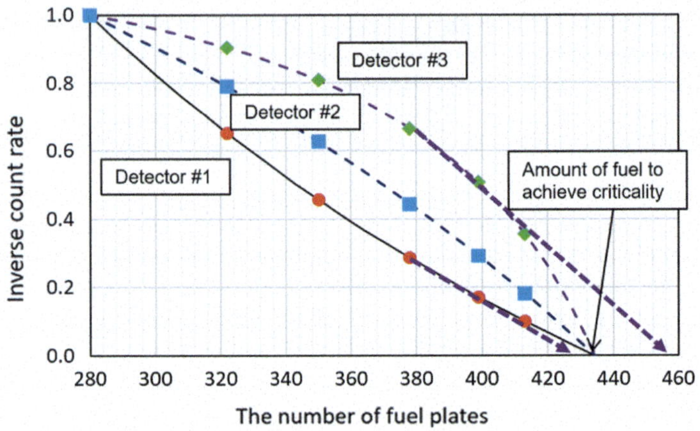

Fig. 3.4 Prediction of the number of fuel plates to attain the criticality through extrapolation in the inverse count rate curve

(8) Confirmation of the critical state and recording of the reactor status data at the critical state

The critical state is confirmed by observing that the reactor power is constant with time (or sustainable fission chain reactions are realized) without any external sources. In other words, the external source is excluded from a core, control rod insertion positions are adjusted, and it is confirmed that the linear range power monitor gives the constant value with time. It can also be confirmed by observing the linear increase in the reactor power when the external neutron source remains in the core. After confirming that the reactor core reaches a critical state, the following state parameters should be recorded:

- The number of the loaded fuel plates,
- Positions of the control rods and the safety rods,
- Water level in the core tank,
- Power level indicated by the linear-scale and log-scale monitors,
- Level of the γ-ray, and
- Temperature of the core tank.

(9) Measurement of the excess reactivity

When a nuclear reactor reaches a critical state with one control rod being partly inserted, it can be said that this reactor is a supercritical system having a positive reactivity which is equal to the reactivity by this control rod insertion. This reactivity over the critical state is known as the excess reactivity. To calculate a minimum critical mass of the reactor, the excess reactivity of the reactor is measured.

In this experiment, the stable period method is used to measure the excess reactivity. The stable period T is defined as a time duration during which the reactor power increases by e times and is calculated as $T = T_{\mathrm{d}}/\ln 2$ where T_{d} is the doubling time during which the value of the linear power monitor becomes doubled. In this method, firstly the reactor power is increased by the control rod withdrawn from a critical reactor. The reactor power increases exponentially, and the excess reactivity is quantified from the increasing rate of the reactor power. This is the fundamental principle of the stable period method. In actual measurement, after the control rod is withdrawn, the doubling time is measured by stopwatches or others, and this doubling time is converted to the stable period and finally the excess reactivity is derived. The theory of the stable period method and how to calculate the excess reactivity will be described in Chap. 4 in detail.

(10) Derivation of the minimum number of fuel plates to achieve criticality

Let the excess reactivity obtained above ρ_{ex} ($\Delta k/k$). In order to derive the minimum number of the fuel plates to achieve criticality, the number of the fuel plates corresponding to the obtained excess reactivity is calculated and this is subtracted from the number of fuel plates loaded in a reactor core when the criticality is achieved. To do this, the value of the reactivity equivalent with the one fuel plate, or the reactivity worth of one fuel plate, should be calculated. The reactivity of one fuel plate is denoted to as α_{plate} ($\Delta k/k \cdot (\text{plate})^{-1}$) here. When α_{plate} is known, the number of the fuel plates corresponding to the excess reactivity ρ_{ex}, N_{ex}, is represented as

$$N_{ex} = \frac{\rho_{ex}}{\alpha_{plate}}, \tag{3.10}$$

and the minimum number of the fuel plates to attain the criticality can be obtained by subtracting this from the number of the fuel plates loaded in a reactor reaching a critical state.

To derive α_{plate}, the measurement data obtained through the control rod calibration experiment are also required in addition to those through the approach-to-criticality experiment. In the control rod calibration experiment, the core excess reactivity is increased by adding fuel plates to the reactor core constructed in the approach-to-criticality experiment due to the reason of the experiment. This will be mentioned in Chap. 4. When the number of the loaded fuel plates and the excess reactivity in the control rod calibration experiment are written as N_{rod} and $\rho_{ex,rod}$, and those in the approach-to-criticality experiment are N_{crit} and $\rho_{ex,crit}$, α_{plate} can be obtained from the following equation:

$$\alpha_{plate} = \frac{\rho_{ex,rod} - \rho_{ex,crit}}{N_{rod} - N_{crit}}. \tag{3.11}$$

As described above, the determination of the minimum number of fuel plates to achieve the criticality is possible after carrying out the control rod calibration experiment.

3.3 Discussions

The report of this experiment should be clearly written and should be easy to understand for readers who have possibility to use the measurement data obtained through the experiment. It is desirable that the readers can reproduce the measurement results with the same procedure and they can understand which conclusions can be drawn from the results. Based on this consideration, the following should be summarized as measurement results:

- Locations of the fuel frames, the control rods, the safety rods, the neutron detectors, and the external neutron source in the reference core.
- Fuel loading information in the reference core and the other cores.
- Prediction of the number of the fuel plates to achieve the criticality at each measurement step.
- Explanations of how to determine the number of the added fuel plates.
- Table of data of the neutron count rates measured by the experimental group participants are belonging to.
- Figure of the inverse count rates for each neutron detector.
- Figure presenting the relation between the water level of the core tank and the neutron count rate of the detector.
- Operation log when a critical state is attained.
- Measurement results of the excess reactivity.
- Minimum number of fuel plates to attain the criticality. After the control rod calibration experiment, the reactivity worth of one fuel plate α_{plate} is derived, and the minimum number of fuel plates to achieve the criticality is calculated.

3.3.1 *Inverse Count Rate*

It is expected that various curves of the inverse count rates dependent on the detector positions are obtained. Discuss this dependence of the shape of the curves on the detector positions by considering the relation in spatial positions among a reactor core, detectors, external source, and the direction of the fuel plate addition. It is also beneficial to compare the curves with those obtained numerically in the preliminary exercises described in Sect. 3.4.2.

3.3.2 *Light Water Reflector Thickness Equivalent to the Infinite Thickness*

Let us remember the change in the neutron count rate when decreasing the water level of the reactor tank from the fully-flooded state through the measurement in Sect. 3.2.1 (5). This change in the neutron count rate is caused by the change in the thickness of the top reflectors; reduction of the reflector thickness reduces the capability of neutron reflection of the water around the top of the reactor core.

Some of neutrons leaked from the reactor core are returned by the reflector and again contribute to the fission chain reaction. When the reflector thickness is increased from zero, this effect is expected to be enhanced. Qualitatively, if the reflector thickness becomes a certain value, the probability of leaked neutrons to reach the reflector with this thickness becomes negligible, and the neutron reflection capability would be saturated. The reflector having this thickness can be regarded as the infinite-thickness reflector. If such phenomena are observed during this experiment, estimate the reflector thickness which is equivalent to the infinite thickness. Furthermore, discuss whether the qualitative explanations on this are possible or not by considering the spatial shapes of neutron flux obtained by the preliminary exercises.

3.3.3 The Number of Fuel Plates to Achieve Criticality

From the difference in the number of the fuel plates to attain the criticality between the predicted value in the preliminary exercises and the actual value obtained through the experiment ΔN, calculate the difference in the unit of reactivity $\Delta\rho$. It can be obtained as

$$\Delta\rho = \Delta N \times \alpha_{\text{plate}}, \tag{3.12}$$

where α_{plate} is the reactivity worth of one fuel plate.

The number of the fuel plates to achieve the criticality predicted from the curves of the inverse count rate is different among the control rod insertion patterns $j = 1, 2$, and 3. This difference in the prediction comes from the difference in the control rod insertion pattern. For example, the prediction at $j = 2$ is for the reactor core in which one control rod is fully inserted and the other two are fully withdrawn, and that at $j = 3$ is for the core in which all these three control rods are fully withdrawn. This difference in the prediction can be explained by the difference in k_{eff} dependent on the control rod insertion pattern. Convert the difference in the predicted number of the fuel plates to achieve the criticality between $j = 1$ or 2 and $j = 3$ to the reactivity by using α_{plate}. Furthermore, compare these reactivities with those obtained through the actual measurement in the control rod calibration experiment considering the detector positions.

3.4 Preliminary Exercises

It is desirable to perform the following preliminary exercises prior to the actual experiment:

- Simulation of the approach-to-criticality experiment with the simple model, and
- Calculation of the amount of nuclear fuels required to make the reactor critical (criticality calculations).

In this preliminary exercises, the amount of the nuclear fuels to make the reactor critical is estimated. This result will be utilized in a comparison with the actual values obtained through the experiment. Nowadays, comparisons between measurement results and simulation results obtained with the high-fidelity advanced computer codes are reflected to the evaluation of the nuclear data such as reaction cross sections in order to improve the accuracy and the reliability of the nuclear data. For accurate prediction of the amount to achieve the criticality, reliable nuclear data and accurate numerical methods based on the neutron transport theory are mandatory, but relatively simple calculation methods based on the neutron diffusion theory are also practical and sufficient to grasp the neutronics property of the reactor core.

In this preliminary exercises, the inverse count rate curve is obtained by simple numerical simulation based on the six-factor formula explained in Sect. 2.3. Then, calculations of the fuel amount to attain the criticality and the neutron flux spatial distribution based on the two-group diffusion theory are carried out.

3.4.1 Reactor Core Model Used in Simulation

To predict the fuel amount to achieve the criticality, the reactor core is simplified into a parallelpiped model shown in Fig. 3.5. It is assumed that the reactor core is surrounded by the light water reflector with the sufficiently large thickness. The coordinate of this model is set as

- Direction of the fuel plate additions as an x-axis,
- Direction corresponding to the column of the fuel frame as the y-axis, and

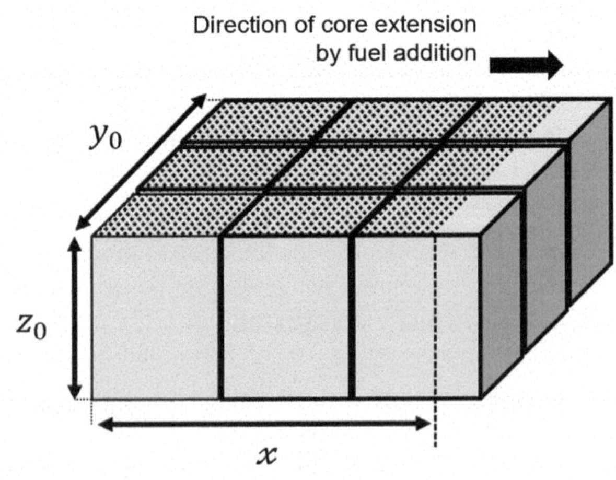

Fig. 3.5 Numerical simulation model and its coordinates

- Axial direction of the fuel plates as the z-axis.

In the approach-to-criticality experiment, the sizes of the reactor core along the y direction (fuel frame column) and the z direction are fixed, so the length of the x-axis in a critical state ($k_{\text{eff}} = 1$) is concerned. The length of the y direction, y_0, is dependent on the reactor core constructed in the actual experiment, and is given as $y_0 = 7.1$ cm × [the number of fuel frame columns]. On the length of the z-direction, $z_0 = 57$ cm, which is an axial length of the fuel meat part, is used by neglecting the aluminum sheath located at the top and bottom of the fuel plates and other structural materials and by assuming that the fuel meat part directly neighbors the light water reflectors.

3.4.2 Simple Simulation of the Approach-to-Criticality Experiment Using the Six-Factor Formula

Simple simulation is carried out for the approach-to-criticality experiment based on the six-factor formula derived in Sect. 2.3:

$$k_{\text{eff}} = \frac{k_\infty e^{-B^2\tau_{\text{T}}}}{1 + L_{\text{T}}^2 B^2}. \tag{3.13}$$

An example of the constants used for this simulation is shown in Table 3.2. These data were obtained by adjusting the four-group constants generated with the computer codes UGMG and THERMOS in order to keep the consistency from the physical point of view of the six-factor formula. The values are dependent on the fuel frame types used in the experiment: C30, C35, or C45. The variables k_∞ and L_{T}^2 are given as

$$k_\infty = \eta\varepsilon p f, \tag{3.14}$$

$$L_{\text{T}}^2 = \frac{D}{\Sigma_{\text{a}}}. \tag{3.15}$$

Next let us discuss the buckling B^2 in the leakage term. Since the six-factor formula is derived for bare nuclear reactors, it is necessary to introduce the proper value of the extrapolation length (or the reflector saving) δ to consider the reflector effect when reactors with neutron reflectors are concerned as shown in Fig. 3.6. More precisely, a reactor with an enhanced size is considered by adding 2δ to the actual dimensions to reduce the neutron leakage rate.

Based on the theory for homogeneous bare thermal reactors, neutron flux spatial distribution of a parallelpiped reactor along the x, y, and z directions are represented as

Table 3.2 An example of the constants used in the six-factor formula in the case of highly-enriched uranium fuels

Core fuel type	C30	C35	C45
τ_T(cm^2)	64.1	56.2	48.4
ηf	1.81	1.75	1.67
p	0.848	0.881	0.915
ε	1.121	1.092	1.064
Σ_a (thermal averaged absorption cross section) (cm^{-1})	0.0930	0.0850	0.0724
D (thermal diffusion coefficient) (cm)	0.271	0.237	0.203
δ (reflector saving) (cm)	9.4	8.2	7.8

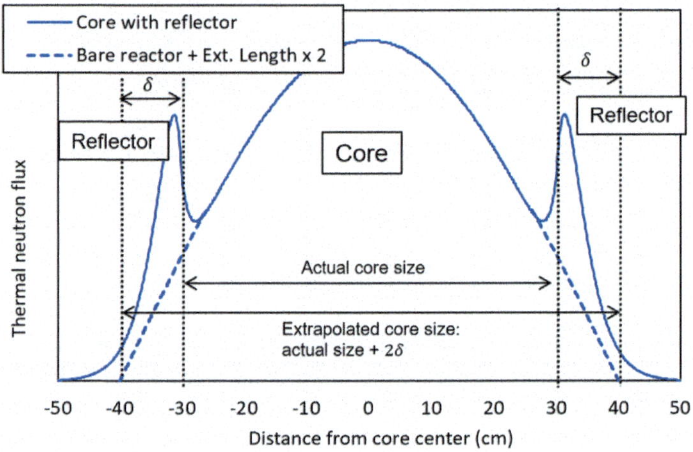

Fig. 3.6 Core size with considering the extrapolation length

$$\phi(x, y, z) = A \cos\left(\frac{\pi}{a}x\right) \cos\left(\frac{\pi}{b}y\right) \cos\left(\frac{\pi}{c}z\right), \tag{3.16}$$

where a, b, and c are sizes of this reactor along each direction, and the buckling is given as

$$B^2 = \left(\frac{\pi}{a}\right)^2 + \left(\frac{\pi}{b}\right)^2 + \left(\frac{\pi}{c}\right)^2. \tag{3.17}$$

In the reactors of the present experiment, the length of the y and z directions are predefined and fixed, and only that of the x direction is unknown. Thus the buckling is a function dependent only on the variable x and is represented as $B^2(x)$. When the length of each direction in the reactor with the consideration of the extrapolation length δ is used, the buckling is written as

$$B^2(x) = \left(\frac{\pi}{x + 2\delta}\right)^2 + \left(\frac{\pi}{y_0 + 2\delta}\right)^2 + \left(\frac{\pi}{z_0 + 2\delta}\right)^2. \tag{3.18}$$

Thus, the effective multiplication factor is also represented as a function on the variable x as

$$k_{\mathrm{eff}}(x) = \frac{k_\infty e^{-B^2(x)\tau_{\mathrm{T}}}}{1 + L_{\mathrm{T}}^2 B^2(x)}. \tag{3.19}$$

Based on the above discussions, the approach-to-criticality experiment is simply simulated; starting with the reference core, the reactor size is enlarged by addition of the nuclear fuel to the reactor core, the curve of the inverse count rate is produced, and the trend of this curve is discussed. In the following simulation, the length along the x direction in the reference core is assumed $x_0 = 15$ cm.

(1) Starting $x = x_0$, increase the variable x gradually and calculate $k_{\mathrm{eff}}(x)$ with the six-factor formula. Then calculate the inverse count rate as $\frac{A_0}{A(x)} = \frac{1-k_{\mathrm{eff}}(x)}{1-k_{\mathrm{eff}}(x_0)}$, and plot it. Note that the numerical results are used in the following item (2) also. The magnitude of the increase in x is determined by observing the shape of the inverse count rate curve.

- Describe the shape of the inverse count rate curve.
- Derive the size of the x direction when the criticality is achieved. In addition, discuss the difference in the prediction by the extrapolation when using the data at the states close to the criticality and those far from it.

(2) Neutron count rates by detectors should include the component of neutrons multiplied in the reactor core and that directly from the external neutron source. Here the former at the core whose size of the x-direction is x is denoted as $A_{\mathrm{core}}(x)$ and the following relation holds:

$$\frac{1 - k_{\mathrm{eff}}(x)}{1 - k_{\mathrm{eff}}(x_0)} = \frac{A_{\mathrm{core}}(x_0)}{A_{\mathrm{core}}(x)}. \tag{3.20}$$

If the constant component of the direct contribution from the external source is represented as A_{s}, the inverse count rate is given as

$$\begin{aligned} \frac{A_{\mathrm{core}}(x_0) + A_{\mathrm{s}}}{A_{\mathrm{core}}(x) + A_{\mathrm{s}}} &= \frac{1 + A_{\mathrm{s}}/A_{\mathrm{core}}(x_0)}{A_{\mathrm{core}}(x)/A_{\mathrm{core}}(x_0) + A_{\mathrm{s}}/A_{\mathrm{core}}(x_0)} \\ &= \frac{1 + f}{A_{\mathrm{core}}(x)/A_{\mathrm{core}}(x_0) + f}, \end{aligned} \tag{3.21}$$

where $f = A_{\mathrm{s}}/A_{\mathrm{core}}(x_0)$. Plot the inverse count rate curves with three or four different values of f: a case of $f = 0$ in which no contribution from the external sources exist is mandatory and the maximum of f can be five to ten.

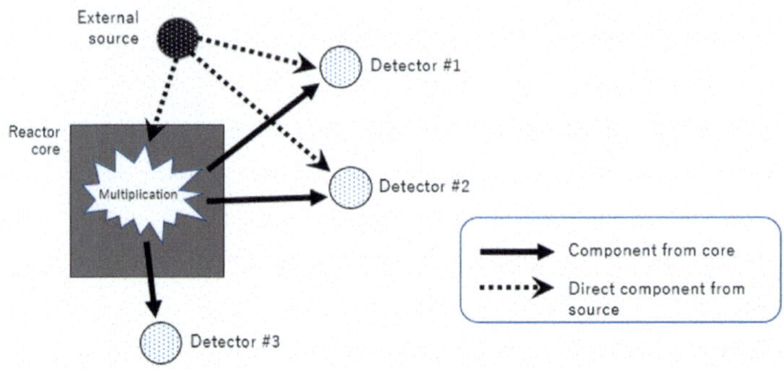

Fig. 3.7 An example of the spatial positions of the external neutron source and neutron detectors

- Observe dependence of the inverse count rate curves on f.
- Discuss this dependence qualitatively.
- If we consider an experimental system as shown in Fig. 3.7, guess the shape of the inverse count rate curves obtained using the measurement data at the detectors 1, 2, and 3.

3.4.3 Prediction of the Reactor Size to Achieve Criticality with the Two-Group Diffusion Theory

Although the six-factor formula is practical to quantitatively evaluate the neutron multiplication in a nuclear reactor, some physical phenomena such as the neutron leakage from a reactor and neutron slowing-down cannot be rigorously taken into account. In the present section, based on the neutron diffusion theory, we discuss the different behaviors of fast and thermal neutrons by treating them separately by dividing the energy range into two with the energy boundary of around 1 eV; fast energy range above the boundary and the thermal range below the boundary. The theoretical details are described in the following Sect. 3.5, separately.

(1) On the given experimental system, derive the size in a critical state using the procedure described in Sect. 3.5. Constants in the two-group diffusion equation are given in Table 3.3.
(2) After the size in a critical state is obtained, calculate the mass of ^{235}U (critical mass) in the core volume of $V = x_0 y_0 z_0$ (cm^3) using the homogenized atomic number densities shown in Table 3.4. Next, by dividing the obtained critical mass by the weight of ^{235}U in one fuel plate (in the case of the HEU fuel) 8.89 g, obtain the number of fuel plates in the critical state.
(3) Referring Sect. 3.5, calculate and plot the fast and thermal neutron flux spatial distributions and a ratio of them. When plotting the neutron flux spatial distributions, proper normalization conditions such as $\phi_{1c}(x = 0) = 1$ should be adopted.

Table 3.3 Example of the constants for two-group diffusion calculations in the case of highly-enriched uranium fuels [unit: (cm) or (cm^{-1})]

	The 1st group (fast group)			The 2nd group (thermal group)			ε	δ
	D_1	Σ_{1a}	$\Sigma_{1\to 2}$	D_2	Σ_{2a}	$\nu\Sigma_{2f}$*		
C30	1.58	0.00320	0.0178	0.271	0.0930	0.168	1.121	9.4
C35	1.54	0.00286	0.0212	0.237	0.0850	0.149	1.092	8.2
C45	1.50	0.00237	0.0254	0.203	0.0724	0.121	1.064	7.8
Water reflector	1.41	0.0	0.0476	0.117	0.0191	0.0	–	–

*$\nu = 2.44$

Table 3.4 Atomic number density of ^{235}U averaged over the reactor core region in the case of the highly-enriched uranium fuel

Fuel type in the core	C30	C35	C45
Atomic number density of ^{235}U (atoms cm^{-3})*	1.867×10^{20}	1.589×10^{20}	1.231×10^{20}

* Averaged over the fuel meat part of 57 cm in the fuel plate

3.5 Application of Two-Group Neutron Diffusion Theory

3.5.1 Two-Group Neutron Diffusion Equation

Let us consider a two-region reactor consisting of a homogeneous parallelpiped core surrounded by neutron reflectors made of light water. In the following, variables for the core and reflector regions are distinguished by the subscripts c and r, and those for the fast and thermal groups also by the subscripts 1 and 2.

Firstly, the neutron balance equation for the fast group in the reactor core region is derived from the neutron diffusion equation in a critical state. Fast neutron production rates at a certain spatial position can be written as $\varepsilon\nu\Sigma_{2fc}\phi_{2c}$ where ν is the averaged number of fast neutrons generated by one fission reaction, Σ_{2fc} is thermal-group fission cross sections, and ε is the fast neutron fission factor. On the other hand, the loss rates of fast-group neutrons can be calculated as a sum of the leakage term $-D_{1c}\nabla^2\phi_{1c}$, the fast neutron absorption term $\Sigma_{1ac}\phi_{1c}$, and the scattering term from the fast group to the thermal group $\Sigma_{1\to 2c}\phi_{1c}$. If the reactor is in a critical state, the neutron generation rates and the loss rates are balanced and the following equation is preserved:

$$D_{1c}\nabla^2\phi_{1c} - (\Sigma_{1ac} + \Sigma_{1\to 2c})\phi_{1c} + \varepsilon\nu\Sigma_{2fc}\phi_{2c} = 0. \quad (3.22)$$

If non-critical states are also considered, the above equation can be generalized by introducing the effective multiplication factor k_{eff} as follows:

$$D_{1c}\nabla^2\phi_{1c} - (\Sigma_{1ac} + \Sigma_{1\to 2c})\phi_{1c} + \frac{\varepsilon\nu\Sigma_{2fc}\phi_{2c}}{k_{eff}} = 0. \tag{3.23}$$

On the thermal-group neutrons in the core region, neutrons scattered from the fast group to the thermal group are balanced with the neutrons lost through the leakage and absorptions, and the following equation is preserved:

$$D_{2c}\nabla^2\phi_{2c} - \Sigma_{2ac}\phi_{2c} + \Sigma_{1\to 2c}\phi_{1c} = 0. \tag{3.24}$$

The fast neutron balance equation in the reflector region can be written as follows without the fission neutron generation term:

$$D_{1r}\nabla^2\phi_{1r} - (\Sigma_{1ar} + \Sigma_{1\to 2r})\phi_{1r} = 0. \tag{3.25}$$

The thermal neutron balance equation in the reflector region can be written as follows like the one for the core region:

$$D_{2r}\nabla^2\phi_{2r} - \Sigma_{2ar}\phi_{2r} + \Sigma_{1\to 2r}\phi_{1r} = 0. \tag{3.26}$$

The above Eqs. (3.22), (3.24), (3.25), and (3.26) are the two-group neutron diffusion equations for this two-region problem. At this moment, the neutron flux is a parameter dependent on three variables *x*, *y*, and *z*.

3.5.2 Analytical Solution to the Two-Region, Two-Group Neutron Diffusion Equation

In this section, the analytical solution to the two-group neutron diffusion equations is derived when the reactor is in a critical state, $k_{eff} = 1$.

Generally, numerical methods such as the finite difference method are employed to solve the simultaneous differential equations, but as described below, the analytical solutions can be obtained by reducing the number of spatial variables to one with the reasonable approximation. In the following, it is assumed that the center position of the reactor core is located at the origin of the coordinate system. Since the homogeneous parallelpiped reactor core is concerned, neutron flux spatial distributions should be symmetric to the xy, yz, and xz planes which cross the origin.

Firstly, we assume that the neutron flux spatial distribution $\phi(x, y, z)$ can be represented with the variable-independent form as $\phi(x, y, z) = \phi(x)\phi(y)\phi(z)$, and neutron flux distribution along the y and z directions can be represented by the cosine function as

$$\phi(x, y, z) = \phi(x)\cos\left(\frac{\pi}{b}y\right)\cos\left(\frac{\pi}{c}z\right), \tag{3.27}$$

where b and c are the length of the y and z directions, respectively. Here we are concerned with the reflected reactor core, so the size of the reactor should be corrected by the extrapolation length δ and the corrected length are defined as follows:

$$b = y_0 + 2\delta, \tag{3.28}$$

$$c = z_0 + 2\delta. \tag{3.29}$$

In this case, the term $\nabla^2\phi$ is represented by

$$\nabla^2\phi = \left[\frac{d^2\phi(x)}{dx^2} - \pi^2\left\{\frac{1}{(y_o + 2\delta)^2} + \frac{1}{(z_o + 2\delta)^2}\right\}\phi(x)\right] \cos\left(\frac{\pi}{y_0 + 2\delta}y\right)\cos\left(\frac{\pi}{z_0 + 2\delta}z\right). \tag{3.30}$$

If the buckling along the horizontal directions (y and z directions) is defined as

$$B_\perp^2 = \pi^2\left\{\frac{1}{(y_0 + 2\delta)^2} + \frac{1}{(z_0 + 2\delta)^2}\right\}, \tag{3.31}$$

Equation (3.30) is rewritten as

$$\nabla^2\phi = \left(\frac{d^2\phi(x)}{dx^2} - B_\perp^2\phi(x)\right)\cos\left(\frac{\pi}{y_0 + 2\delta}y\right)\cos\left(\frac{\pi}{z_0 + 2\delta}z\right). \tag{3.32}$$

After substituting Eq. (3.32) into the two-group diffusion equation and removing the cosine terms of the neutron flux for the y and z directions, the following equations dependent only on the variable x are derived:

$$D_{1c}\frac{d^2\phi_{1c}(x)}{dx^2} - (\Sigma_{1ac} + \Sigma_{1\to 2c} + D_{1c}B_\perp^2)\phi_{1c}(x) + \varepsilon\nu\Sigma_{2fc}\phi_{2c}(x) = 0, \tag{3.33}$$

$$D_{2c}\frac{d^2\phi_{2c}(x)}{dx^2} - (\Sigma_{2ac} + D_{2c}B_\perp^2)\phi_{2c}(x) + \Sigma_{1\to 2c}\phi_{1c}(x) = 0, \tag{3.34}$$

$$D_{1r}\frac{d^2\phi_{1r}(x)}{dx^2} - (\Sigma_{1ar} + \Sigma_{1\to 2r} + D_{1r}B_\perp^2)\phi_{1r}(x) = 0, \tag{3.35}$$

$$D_{2r}\frac{d^2\phi_{2r}(x)}{dx^2} - (\Sigma_{2ar} + D_{2r}B_\perp^2)\phi_{2r}(x) + \Sigma_{1\to 2r}\phi_{1r}(x) = 0. \tag{3.36}$$

These equations are different from the original two-group diffusion Eqs. (3.22), (3.24), (3.25), and (3.26) for the three-dimensional space in that these are for a one-dimensional problem dependent only on x. Furthermore, the term $DB_{\perp}^2$ is added to the second term in the left-hand side. This term represents the neutron leakage along the y and z directions, and is regarded as the fictitious absorption. By introducing the buckling perpendicular to the concerned direction, the x direction at present, neutron leakage along the directions which are not explicitly considered is treated as neutron losses by the absorption.

In the following, the simultaneous Eqs. (3.33) to (3.36) are solved in the core region and the reflector region, respectively.

3.5.2.1 Neutron Flux in the Reactor Core Region

In order to simplify the equations, the following parameters are newly defined:

$$\Sigma_{1c} = \Sigma_{1ac} + \Sigma_{1\to 2c} + D_{1c}B_{\perp}^2, \tag{3.37}$$

$$\Sigma_{2c} = \Sigma_{2ac} + D_{2c}B_{\perp}^2. \tag{3.38}$$

The cross sections Σ_{1c} and Σ_{2c} are removal cross sections from the first and second groups by the absorption, the scattering to the lower group, and the leakage.

Here we assume that the neutron flux distributions along the x direction, $\phi_{1c}(x)$ and $\phi_{2c}(x)$, satisfy the following equations, respectively:

$$\frac{d^2\phi_{1c}(x)}{dx^2} + B_x^2\phi_{1c}(x) = 0, \tag{3.39}$$

$$\frac{d^2\phi_{2c}(x)}{dx^2} + B_x^2\phi_{2c}(x) = 0. \tag{3.40}$$

Using them, Eqs. (3.33) and (3.34) can be rewritten, respectively, as

$$-D_{1c}B_x^2\phi_{1c}(x) - \Sigma_{1c}\phi_{1c}(x) + \varepsilon\nu\Sigma_{2fc}\phi_{2c}(x) = 0, \tag{3.41}$$

$$-D_{2c}B_x^2\phi_{2c}(x) - \Sigma_{2c}\phi_{2c}(x) + \Sigma_{1\to 2c}\phi_{1c}(x) = 0. \tag{3.42}$$

These Eqs. (3.41) and (3.42) can be represented in a matrix form as

$$\begin{pmatrix} -D_{1c}B_x^2 - \Sigma_{1c} & \varepsilon\nu\Sigma_{2fc} \\ \Sigma_{1\to 2c} & -D_{2c}B_x^2 - \Sigma_{2c} \end{pmatrix}\begin{pmatrix} \phi_{1c}(x) \\ \phi_{2c}(x) \end{pmatrix} = \begin{pmatrix} 0 \\ 0 \end{pmatrix}. \tag{3.43}$$

This simultaneous equation has the non-trivial solution (neutron flux is non-zero) if the determinant of the matrix in the left-hand side is zero:

$$(-D_{1c}B_x^2 - \Sigma_{1c})(-D_{2c}B_x^2 - \Sigma_{2c}) - \varepsilon\nu\Sigma_{2fc}\Sigma_{1\to 2c} = 0. \tag{3.44}$$

This equation leads to the following second-order polynomial equation on B_x^2 as

$$D_{1c}D_{2c}(B_x^2)^2 + (D_{1c}\Sigma_{2c} + D_{2c}\Sigma_{1c})B_x^2 + (\Sigma_{1c}\Sigma_{2c} - \varepsilon\nu\Sigma_{2fc}\Sigma_{1\to 2c}) = 0. \tag{3.45}$$

If the solutions of B_x^2 are written as μ^2 and $-\lambda^2$, we can obtain the following equations from the formula of the solutions to the second-order polynomial equation:

$$\mu^2 = \frac{1}{2D_{1c}D_{2c}}\Big[-(D_{1c}\Sigma_{2c} + D_{2c}\Sigma_{1c}), \\ +\sqrt{(D_{1c}\Sigma_{2c} + D_{2c}\Sigma_{1c})^2 - 4D_{1c}D_{2c}(\Sigma_{1c}\Sigma_{2c} - \varepsilon\nu\Sigma_{2fc}\Sigma_{1\to 2c})}\Big] \tag{3.46}$$

$$-\lambda^2 = \frac{1}{2D_{1c}D_{2c}}\Big[-(D_{1c}\Sigma_{2c} + D_{2c}\Sigma_{1c}) \\ -\sqrt{(D_{1c}\Sigma_{2c} + D_{2c}\Sigma_{1c})^2 - 4D_{1c}D_{2c}(\Sigma_{1c}\Sigma_{2c} - \varepsilon\nu\Sigma_{2fc}\Sigma_{1\to 2c})}\Big]. \tag{3.47}$$

Thus the general solutions of $\phi_{1c}(x)$ and $\phi_{2c}(x)$ can be represented as the linear combination of $X(x)$ and $Y(x)$ which satisfy the following differential equations:

$$\frac{d^2X(x)}{dx^2} + \mu^2 X(x) = 0, \tag{3.48}$$

$$\frac{d^2Y(x)}{dx^2} - \lambda^2 Y(x) = 0. \tag{3.49}$$

By adopting the symmetric condition of $X(x)$ and $Y(x)$ with respect to $x = 0$, the following equations are derived:

$$X(x) = \cos(\mu x), \tag{3.50}$$

$$Y(x) = \cosh(\lambda x). \tag{3.51}$$

Regarding $X(x)$, substitutions of $\phi_{1c}(x) = AX(x)$ and $\phi_{2c}(x) = A'X(x)$ to Eq. (3.42), the following equation can be derived. Note that B_x^2 in Eq. (3.42) is replaced by μ^2:

$$S_1 \equiv \frac{A'}{A} = \frac{\Sigma_{1\to 2c}}{D_{2c}\mu^2 + \Sigma_{2c}}. \tag{3.52}$$

Similary, substitutions of $\phi_{1c}(x) = CY(x)$ and $\phi_{2c}(x) = C'Y(x)$ to Eq. (3.42), the following equation can be derived. In this case, B_x^2 is replaced by $-\lambda^2$:

$$S_2 \equiv \frac{C'}{C} = \frac{\Sigma_{1\to 2c}}{-D_{2c}\lambda^2 + \Sigma_{2c}}. \tag{3.53}$$

Based on the above discussions, the following equations can be obtained:

$$\phi_{1c}(x) = A\cos(\mu x) + C\cosh(\lambda x), \tag{3.54}$$

$$\phi_{2c}(x) = AS_1\cos(\mu x) + CS_2\cosh(\lambda x). \tag{3.55}$$

The parameters S_1 and S_2 defined above are called the coupling coefficients since they represent the relations between the neutron fluxes of the first and second groups.

3.5.2.2 Neutron Flux in the Reflector Region

Next, we derive the neutron flux spatial distribution in the reflector region. With the same procedure as the core region, the following parameters are introduced for simplicity:

$$\Sigma_{1r} = \Sigma_{1ar} + \Sigma_{1\to 2r} + D_{1r}B_\perp^2, \tag{3.56}$$

$$\Sigma_{2r} = \Sigma_{2ar} + D_{2r}B_\perp^2. \tag{3.57}$$

The diffusion equation for the first group

$$D_{1r}\frac{d^2\phi_{1r}(x)}{dx^2} - \Sigma_{1r}\phi_{1r}(x) = 0, \tag{3.58}$$

is rewritten as

$$\frac{d^2\phi_{1r}(x)}{dx^2} - \kappa_{1r}^2\phi_{1r}(x) = 0, \tag{3.59}$$

where $\kappa_{1r}^2 = \frac{\Sigma_{1r}}{D_{1r}}$. Among the solutions to Eq. (3.58), one which becomes zero in $x \to \infty$ is represented as

$$\varphi_{1r}(x) = Fe^{-\kappa_{1r}|x|}, \tag{3.60}$$

where F is an arbitrary constant.

Similarly, the diffusion equation for the second group

$$D_{2r}\frac{d^2\phi_{2r}(x)}{dx^2} - \Sigma_{2r}\phi_{2r}(x) + \Sigma_{1\to 2r}\phi_{1r}(x) = 0, \tag{3.61}$$

is rewritten as

$$\frac{d^2\phi_{2r}(x)}{dx^2} - \kappa_{2r}^2\phi_{2r}(x) + \frac{\Sigma_{1\to 2r}}{D_{2r}}\phi_{1r}(x) = 0, \tag{3.62}$$

where $\kappa_{2r}^2 = \frac{\Sigma_{2r}}{D_{2r}}$. By assuming that the solution to this equation can be represented as

$$\phi_{2r}(x) = \hat{F}e^{-\kappa_{1r}|x|} + Ge^{-\kappa_{2r}|x|}, \tag{3.63}$$

and substituting this into Eq. (3.62), the following equation is derived:

$$\left(\kappa_{1r}^2\hat{F} - \kappa_{2r}^2\hat{F} + \frac{\Sigma_{1\to 2r}}{D_{2r}}F\right)e^{-\kappa_{1r}|x|} = 0. \tag{3.64}$$

This Eq. (3.64) is preserved with the arbitrary value of x if the following equation is satisfied:

$$\kappa_{1r}^2\hat{F} - \kappa_{2r}^2\hat{F} + \frac{\Sigma_{1\to 2r}}{D_{2r}}F = 0. \tag{3.65}$$

This leads to

$$S_3 \equiv \frac{\hat{F}}{F} = \frac{\frac{\Sigma_{1\to 2r}}{D_{2r}}}{\kappa_{2r}^2 - \kappa_{1r}^2}. \tag{3.66}$$

The parameter S_3 is the third coupling coefficient and couples the first- and second-group neutron fluxes with each other in the reflector region.

3.5.2.3 Matrix Equation for Criticality

As a conclusion, neutron flux spatial distributions in the core and reflector regions are represented as follows:

$$\phi_{1c}(x) = A\cos(\mu x) + C\cosh(\lambda x), \tag{3.67}$$

$$\phi_{2c}(x) = S_1 A\cos(\mu x) + S_2 C\cosh(\lambda x), \tag{3.68}$$

$$\phi_{1r}(x) = Fe^{-\kappa_{1r}|x|}, \tag{3.69}$$

$$\phi_{2r}(x) = S_3 Fe^{-\kappa_{1r}|x|} + Ge^{-\kappa_{2r}|x|}. \tag{3.70}$$

Next let us consider the boundary conditions for Eqs. (3.67) to (3.70); the neutron flux and the neutron current $J = -Dd\phi/dx$ of the first and second groups should be continuous on the boundary between the core and reflector regions. For simplicity, the core size in a critical state is assumed $2a$ and the first-order differential of the neutron flux ϕ is represented as ϕ'. The boundary condition can be written as follows:

$$\phi_{1c}(a) = \phi_{1r}(a), \tag{3.71}$$

$$-D_{1c}\phi'_{1c}(a) = -D_{1r}\phi'_{1r}(a), \tag{3.72}$$

$$\phi_{2c}(a) = \phi_{2r}(a), \tag{3.73}$$

$$-D_{2c}\phi'_{2c}(a) = -D_{2r}\phi'_{2r}(a). \tag{3.74}$$

From the above equations, the following set of equations can be obtained:

$$A\cos(\mu a) + C\cosh(\lambda a) = Fe^{-\kappa_{1r}a}, \tag{3.75}$$

$$-AD_{1c}\mu\sin(\mu a) + CD_{1c}\lambda\sinh(\lambda a) = -FD_{1r}\kappa_{1r}e^{-\kappa_{1r}a}, \tag{3.76}$$

$$S_1 A\cos(\mu a) + S_2 C\cosh(\lambda a) = S_3 Fe^{-\kappa_{1r}a} + Ge^{-\kappa_{2r}a}, \tag{3.77}$$

$$\begin{aligned} &- S_1 AD_{2c}\mu\sin(\mu a) + S_2 CD_{2c}\lambda\sinh(\lambda a) \\ &= -S_3 FD_{2r}\kappa_{1r}e^{-\kappa_{1r}a} - GD_{2r}\kappa_{2r}e^{-\kappa_{2r}a}. \end{aligned} \tag{3.78}$$

Unknown parameters A, C, F, and G are treated as entries of a column vector, and then Eqs. (3.75) to (3.78) is written in a matrix form as

$$\begin{pmatrix} \cos(\mu a) & \cosh(\lambda a) & -e^{-\kappa_{1r}a} & 0 \\ -D_{1c}\mu\sin(\mu a) & D_{1c}\lambda\sinh(\lambda a) & D_{1r}\kappa_{1r}e^{-\kappa_{1r}a} & 0 \\ S_1\cos(\mu a) & S_2\cosh(\lambda a) & -S_3 e^{-\kappa_{1r}a} & -e^{-\kappa_{2r}a} \\ -S_1 D_{2c}\mu\sin(\mu a) & S_2 D_{2c}\lambda\sinh(\lambda a) & S_3 D_{2r}\kappa_{1r}e^{-\kappa_{1r}a} & D_{2r}\kappa_{2r}e^{-\kappa_{2r}a} \end{pmatrix} \begin{pmatrix} A \\ C \\ F \\ G \end{pmatrix} = 0 \tag{3.79}$$

If non-trivial solution to Eq. (3.79) exists, the determinant of the matrix in the left-hand side should be zero:

$$\begin{vmatrix} \cos(\mu a) & \cosh(\lambda a) & -e^{-\kappa_{1r}a} & 0 \\ -D_{1c}\mu\sin(\mu a) & D_{1c}\lambda\sinh(\lambda a) & D_{1r}\kappa_{1r}e^{-\kappa_{1r}a} & 0 \\ S_1\cos(\mu a) & S_2\cosh(\lambda a) & -S_3e^{-\kappa_{1r}a} & -e^{-\kappa_{2r}a} \\ -S_1D_{2c}\mu\sin(\mu a) & S_2D_{2c}\lambda\sinh(\lambda a) & S_3D_{2r}\kappa_{1r}e^{-\kappa_{1r}a} & D_{2r}\kappa_{2r}e^{-\kappa_{2r}a} \end{vmatrix} = 0 \quad (3.80)$$

This equation is sometimes called the matrix equation for criticality. This equation can be rewritten as

$$\begin{vmatrix} 1 & 1 & 1 & 0 \\ -D_{1c}\mu\tan(\mu a) & D_{1c}\lambda\tanh(\lambda a) & -D_{1r}\kappa_{1r} & 0 \\ S_1 & S_2 & S_3 & 1 \\ -S_1D_{2c}\mu\tan(\mu a) & S_2D_{2c}\lambda\tanh(\lambda a) & -S_3D_{2r}\kappa_{1r} & -D_{2r}\kappa_{2r} \end{vmatrix} = 0, \quad (3.81)$$

and finally the following equation can be derived:

$$\mu\tan(\mu a) = \frac{[-D_{1r}D_{2c}S_2\kappa_{1r} + D_{1c}D_{2r}\{\kappa_{2r}(S_1 - S_3) + \kappa_{1r}S_3\}]\lambda\tanh(\lambda a) + D_{1r}D_{2r}(S_1 - S_2)\kappa_{1r}\kappa_{2r}}{D_{1c}D_{2c}(S_1 - S_2)\lambda\tanh(\lambda a) + D_{1r}D_{2c}S_1\kappa_{1r} + D_{1c}D_{2r}\{S_3(\kappa_{2r} - \kappa_{1r}) - S_2\kappa_{2r}\}}. \quad (3.82)$$

Since an analytical value for a cannot be obtained from Eq. (3.82), numerical calculations are generally employed to obtain a and the critical size $2a$.

There is a more practical procedure to roughly estimate a. It can be easily found that the value of λ is approximately 0.6 if the constants given in Table 3.3 are substituted to Eq. (3.47). In addition, if we use the half of the critical size obtained in the item (1) of Sect. 3.4.2 with the simple model for the approach-to-criticality experiment, we can find that the approximation of $\tanh(\lambda a) \approx 1$ is reasonable. This can be confirmed by actually substituting the values. Based on the above, Eq. (3.82) is rewritten as

$$\mu\tan(\mu a) = \frac{[-D_{1r}D_{2c}S_2\kappa_{1r} + D_{1c}D_{2r}\{\kappa_{2r}(S_1 - S_3) + \kappa_{1r}S_3\}]\lambda + D_{1r}D_{2r}(S_1 - S_2)\kappa_{1r}\kappa_{2r}}{D_{1c}D_{2c}(S_1 - S_2)\lambda + D_{1r}D_{2c}S_1\kappa_{1r} + D_{1c}D_{2r}\{S_3(\kappa_{2r} - \kappa_{1r}) - S_2\kappa_{2r}\}}, \quad (3.83)$$

and an approximate for a can be represented as

$$a = \frac{1}{\mu}\arctan\left(\frac{1}{\mu}\frac{[-D_{1r}D_{2c}S_2\kappa_{1r} + D_{1c}D_{2r}\{\kappa_{2r}(S_1 - S_3) + \kappa_{1r}S_3\}]\lambda + D_{1r}D_{2r}(S_1 - S_2)\kappa_{1r}\kappa_{2r}}{D_{1c}D_{2c}(S_1 - S_2)\lambda + D_{1r}D_{2c}S_1\kappa_{1r} + D_{1c}D_{2r}\{S_3(\kappa_{2r} - \kappa_{1r}) - S_2\kappa_{2r}\}}\right). \quad (3.84)$$

This approximate solution can be obtained without any complicated numerical calculations, so this can be used as an initial guess when solving Eq. (3.82) numerically.

3.5.2.4 Neutron Flux Spatial Distribution

In order to obtain the spatial distribution of the fast and thermal neutron flux spatial distributions, it is necessary to determine the values of the constants A to G in Eqs. (3.67) to (3.70). When the reactor is in a critical state, however, the matrix equation for the criticality is preserved, and Eqs. (3.75) to (3.78) are not independent of each other, and the constants A to G cannot be uniquely determined. Thus, one of these four constants is treated as unknown, and the other three constants are represented by the unknown constant. Here, let the constant A unknown, and attempt to obtain the other three constants. Using Eqs. (3.75) and (3.76), the constant F can be dropped as

$$C = -A\frac{D_{1r}\kappa_{1r}\cos(\mu a) - D_{1c}\mu\sin(\mu a)}{D_{1r}\kappa_{1r}\cosh(\lambda a) + D_{1c}\lambda\sinh(\lambda a)}. \tag{3.85}$$

Through properly manipulating Eqs. (3.75) to (3.78), the constants F and G can be also represented by the constant A.

Chapter 4
Control Rod Calibration Experiment

Abstract To assure the safety of a nuclear reactor during operation, it is important to know the control rod worth, which is an absolute value of negative reactivity induced by a control rod insertion. The measurement for the control rod worth is known as the control rod calibration. In addition to the safety viewpoints, the control rod calibration is important also in conducting other reactor physics experiments since various reactivity effects can be quantified from the known control rod worth. To measure the control rod worth during this experimental program at KUCA, the stable period method, the compensation method, and the rod drop method are employed. Details of these methods are described in this chapter.

Keywords Control rod worth · Calibration · Stable period method · Rod drop method · Compensation method · First-order perturbation theory

4.1 Objectives

In order to keep a critical state or to change a power of a nuclear reactor, fission chain reaction should be controlled somehow, and one of these mechanisms equipped in a nuclear reactor is a control rod. Control rods are generally made of neutron absorbing materials, such as boron (B), cadmium (Cd), hafnium (Hf), etc. When a control rod made of neutron absorbing materials is inserted into a reactor core, fission chain reaction is restricted by neutron absorptions, and when it is withdrawn from a core, fission chain reaction is enhanced.[1] Control rods are used to change the effective multiplication factor k_{eff}, and in the case of zero-power reactors like

[1] IN some specific reactors such as an educational reactor ANG-201 K in Kyung Hee University, Korea, using nuclear fuel itself as a control rod, positive reactivity is added when control rods are inserted. There are other mechanisms to control the fission chain reaction than uses of control rods. In some reactors including KUCA, especially critical assemblies, a reactivity control method not of using any neutron absorbing materials but of adjusting the neutron leakage rate from a reactor core through changing the amount of moderator material is sometimes adopted. In PWRs, reactivity control is performed also by changing the density of boric acid (H_3BO_3) in moderator in addition to the control rods.

C. H. Pyeon et al., *Reactor Laboratory Experiments at Kyoto University Critical Assembly*, https://doi.org/10.1007/978-981-97-8070-9_4

KUCA, a control rod position in a critical state ($k_{\mathrm{eff}} = 1$) is unchanged since there is no reactivity feedback due to the temperature changes.[2]

To assure the safety of a nuclear reactor during operation, it is important to know the control rod reactivity worth and reactivity insertion rate per unit time. The former is an absolute value of negative reactivity induced by a control rod full insertion from a fully-withdrawn state. In addition, to assure the safety of a reactor in a shutdown state, it is also important to grasp the shutdown margin, which is an absolute value of negative reactivity induced by full insertions of all control and safety rods. From the safety point of view, reactor restrictions are given to these parameters according to a reactor type, and it is required to be confirmed during an inspection at termination of a reactor commission and regular inspections that these reactor restrictions are satisfied. The measurement for these safety parameters relevant to control rods is known as the control rod calibration. The control rod calibration is generally carried out after a reactor core is set up through the approach-to-criticality experiment.

In addition to these safety viewpoints, the control rod calibration is important also in conducting other reactor physics experiments. For example, when reactivity effects such as temperature effects, moderator density (or voiding) effects, neutron absorbing material insertion effects, etc. are measured in a critical state, these reactivity effects are compensated by the control rod movement so as to maintain the critical state. By doing this, the target reactivity being measured can be quantified from the known control rod reactivity. If the reactivity effect induced by the control rod movement is measured in advance as a scaler of the reactivity effect, various reactivity effects other than the control rod worth can be quantified through the measurements.

The reactivity effect induced by the control rod movement mentioned above can be called the integral reactivity worth, and it is defined as a reactivity induced by the control rod movement from a certain position to another position. Thus, the total reactivity worth due to the control rod insertion from a fully-withdrawn state to a fully-inserted state is interpreted as an integral reactivity from a fully-withdrawn position to a fully-inserted position. On the other hand, it is convenient to quantify the reactivity inserted by the slight movement of control rod at every position. This reactivity is called the differential reactivity worth and is defined as reactivity induced by control rod movement of unit length. As indicated by their names, differentiating an integral reactivity at a certain position gives a differential reactivity at this position, and integrating a differential reactivity in a certain range gives an integral reactivity in this range. If these integral (differential) reactivities are plotted in a figure with x-axis of the control rod insertion position, these are called a integral (differential) reactivity curve.

[2] Let us consider that a positive reactivity is inserted to a power reactor core. The reactor power and heat energy should increase, and finally boiling of coolant starts. In general nuclear reactors, this results in the negative reactivity insertion and the reactor power increase is terminated. Like this example, reactivity change due to the status change caused by the other reactivity change is known as the reactivity feedback. Usual nuclear reactors are designed so as to have the negative reactivity feedback mechanism.

The following methods can be used to measure the control rod worth:

(1) the stable period method,
(2) the compensation method (or the standard reactivity method),
(3) the rod drop method,
(4) the subcriticality methods such as the neutron source multiplication method and the pulsed neutron source method, and
(5) the inverse kinetics method.

In the experiment at KUCA, the stable period method and the compensation method which can be foundation of the measurement of various types of reactivity, and the rod drop method, which is used to measure the relatively large negative reactivity, are employed.

4.2 Principle

4.2.1 *Stable Period Method*

Based on the six-group delayed neutron precursor model and the point kinetics equation, a change in a reactor power with time is described as a linear combination of seven exponential functions having the exponent $\omega_j t$ as described in Sect. 2.6.5. When a step-type positive reactivity ($0 < \rho < 1$ dollar) is inserted to a critical reactor, only one among ω_j takes a positive value. If it is denoted to as ω_1, the other ω_j s are negative, so the reactor power finally increases with time according to

$$n(t) \sim e^{\omega_1 t}. \tag{4.1}$$

On the other hand, if a step-type negative reactivity is inserted to a reactor core, all of ω_j s become negative as described in Sect. 2.6.5. In this case, since the term corresponding to ω_1, which is the absolutely smallest one among ω_j s, decreases more slowly than the others, a reactor power finally decreases with time according to Eq. (4.1) like a positive reactivity insertion. The inverse of ω_1 in the above discussion

$$T = \frac{1}{\omega_1} \tag{4.2}$$

is known as the stable period, and is defined as a duration for which a reactor power becomes by e times or $1/e$ times.

By substituting Eq. (4.2) into Eq. (2.62), the following equation relating the reactivity inserted to a reactor core ρ to a stable period T is derived:

$$\rho = \frac{\ell}{T + \ell} + \frac{T}{T + \ell} \sum_{i=1}^{6} \frac{\beta_{\mathrm{eff},i}}{1 + \lambda_i T}. \tag{4.3}$$

This suggests that when a step-type reactivity ρ is inserted to a reactor core, this reactivity can be quantified with Eq. (4.3) from a measurement result of a stable period T. A method of quantifying an inserted reactivity from a measured reactor period T is known as the stable period method. In experiments at KUCA, magnitude of an inserted reactivity is determined so as to make a stable period roughly from 30 to 100 s to consider the safety issues and to ease the measurement. At KUCA, an alarm is activated when a period becomes 30 s; all control rods are simultaneously inserted (or a reactor is auto run-down) when a period becomes 15 s; a scrum signal occurs when a period becomes 10 s.

The prompt neutron lifetime l which can be found in Eq. (4.3) is shorter than 10^{-3} s in every type of reactors as shown in Sect. 2.6.3. Thus, if the measurement is carried out with a period of longer than 10 s, the first term of the right-hand side in Eq. (4.3) becomes smaller than 10^{-4} except heavy water- or graphite-moderated reactors. On the other hand, we can roughly estimate the second term of $i = 4$ as 7×10^{-4} when we use the values $\beta_4 = 2.57 \times 10^{-3}$, $\beta_{\mathrm{eff},4} = \beta_4 \times 1.10$, and $\lambda_4 = 0.301\ \mathrm{s}^{-1}$ according to Ref. [1]. The second term is determined as a sum of all these terms on the group i, and we can estimate the second term as approximately 10^{-3}. Thus, we can assume that the first term in Eq. (4.3) is negligibly smaller than the second term, and then the reactivity can be simply calculated with the following equation:

$$\rho = \sum_{i=1}^{6} \frac{\beta_{\mathrm{eff},i}}{1 + \lambda_i T}. \tag{4.4}$$

Values of delayed neutron precursor-wise effective delayed neutron fractions $\beta_{\mathrm{eff},i}$ are dependent on nuclide composition of the fuel, core size and structure, neutron energy spectra, etc., even though the same fissile uranium-235 (^{235}U) is used as a fuel. Generally numerical simulation is required to obtain them whereas several experimental methods to obtain β_{eff} have been proposed so far. Due to this, when a stable period is measured in a given reactor core, it may happen that the values of $\beta_{\mathrm{eff},i}$ for this reactor core are not yet determined. Even in such cases, it is desired that the reactivity is deduced from the stable period somehow. Precursor group-wise delayed neutron fraction $\beta_{\mathrm{eff},i}$ can be represented using the precursor group-dependent constant γ_i as

$$\beta_{\mathrm{eff},i} = \gamma_i \beta_i, \tag{4.5}$$

and this is approximated as follows using the precursor group-independent constant γ [2]:

$$\beta_{\mathrm{eff},i} = \gamma \beta_i. \tag{4.6}$$

With this equation, the following equation can be derived with the total delayed neutron fraction β:

Table 4.1 Calculation example of β_{eff} and l in the case of highly-enriched uranium fuel

Core name	Prompt neutron lifetime l (second)	Effective delayed neutron fraction β_{eff}	γ
C45G0	5.52×10^{-5}	7.537×10^{-3}	1.178
C35G0	4.92×10^{-5}	7.611×10^{-3}	1.189
C30G0	4.79×10^{-5}	7.673×10^{-3}	1.199

$$\beta_{eff} \equiv \sum_i \beta_{eff,i} = \gamma \sum_i \beta_i = \gamma\beta. \tag{4.7}$$

Furthermore, if Eq. (4.6) is divided by Eq. (4.7) side by side, $\beta_{eff,i}/\beta_{eff}$ can be approximated with the constant β_i/β which is independent of the reactor core [2]:

$$\frac{\beta_{eff,i}}{\beta_{eff}} = \frac{\beta_i}{\beta} = a_i. \tag{4.8}$$

Then, the following equation is derived by transforming Eq. (4.4):

$$\frac{\rho}{\beta_{eff}} = \sum_{i=1}^{6} \frac{a_i}{1 + \lambda_i T}. \tag{4.9}$$

As described above, the left-hand side of Eq. (4.9) is a reactivity in the dollar unit, so now we can find that the reactivity in the dollar unit can be quantified from the measured stable period T and the known physical parameters a_i and λ_i. Values for these parameters in ^{235}U fission by thermal neutrons can be cited from Table 2. 1. An example of β_{eff} and the prompt neutron lifetime l of the KUCA-C core with highly-enriched uranium fuels is shown in Table 4.1.

4.2.2 Rod Drop Method

The reactor shutdown margin is defined as how deeply a reactor is in a subcritical state when all control rods are fully inserted. This shutdown margin should be significantly large in comparison with reactivities which are treated in experiments with the conventional stable period method to assure the reactor safety. Furthermore, negative reactivity worth induced by a full insertion of each of control rods is also important to assure the reactor safety during the normal operation. These large negative reactivities can be measured by the rod drop method easily. When the subcriticality is shallow (absolute value of negative reactivity is small), the neutron source multiplication method using the external source is often used. Please refer to Sect. 10.1.

There are two data processing methods for the rod drop method: the extrapolation method and the integral method. In the following, each of them is described.

4.2.2.1 Extrapolation Method

Let us consider the point kinetics Eqs. (2.54) and (2.55). If we assume that a reactor power in a critical state is kept constant for longer time than the lifetime of delayed neutron precursors, a precursor density C_{i0} should satisfy $\frac{dC_{i0}}{dt} = 0$, and the following equation can be derived:

$$\sum_i \lambda_i C_{i0} = \sum_i \frac{\beta_{i,\mathrm{eff}}}{\Lambda} n_0 = \frac{\beta_{\mathrm{eff}}}{\Lambda} n_0, \tag{4.10}$$

where n_0 is the neutron density in this stationary state.

Let us assume that a reactivity is changed from zero to ρ by a control rod immediate insertion into a critical reactor. Just after the control rod insertion, the delayed neutron precursor densities are unchanged from those in the critical state since the precursor density should be significantly larger than the neutron density in the state before the control rod insertion and the change in the precursor density after the control rod insertion should be negligible. Thus, Eq. (2.54) can be written as follows:

$$\frac{dn(t)}{dt} = \frac{\rho - \beta_{\mathrm{eff}}}{\Lambda} n(t) + \frac{\beta_{\mathrm{eff}}}{\Lambda} n_0. \tag{4.11}$$

The solution to this differential equation can be obtained as follows using the initial condition $n(0) = n_0$:

$$n(t) = \frac{\beta_{\mathrm{eff}} n_0}{\beta_{\mathrm{eff}} - \rho} - \frac{\rho n_0}{\beta_{\mathrm{eff}} - \rho} e^{-\frac{\beta_{\mathrm{eff}} - \rho}{\Lambda} t} \tag{4.12}$$

Since the second term in the right-hand side immediately decreases with the period of $\Lambda / (\beta_{\mathrm{eff}} + |\rho|)$ when $\rho < 0$, the reactor power also immediately becomes

$$n_1 = \frac{\beta_{\mathrm{eff}}}{\beta_{\mathrm{eff}} - \rho} n_0. \tag{4.13}$$

Thus, from the reactor power levels before and after the control rod insertion, n_0 and n_1, respectively, the reactivity can be obtained as

$$\rho = \frac{n_1 - n_0}{n_1} \beta_{\mathrm{eff}}. \tag{4.14}$$

Different from Eq. (4.9) in the stable period method, Eq. (4.14) does not include the physical constants a_i and λ_i, so reactivity in the dollar unit can be deduced with

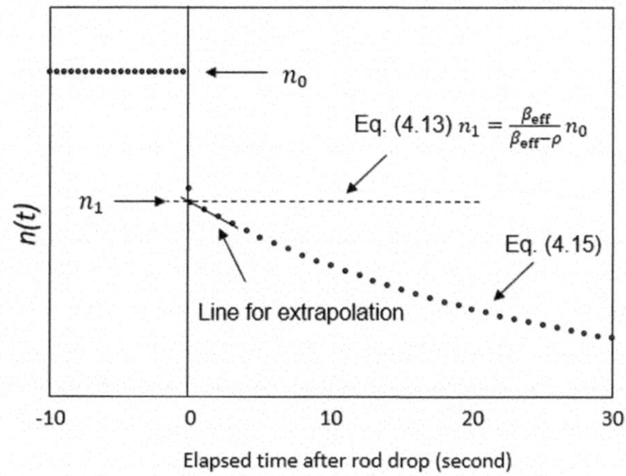

Fig. 4.1 Power change after the rod drop

this method even though these physical constants are unknown. Actually, there are decay reactions of the precursors, so the reactor power is not constant and decreases with time as shown in Fig. 4.1. To consider this, it is necessary to obtain n_1 by extrapolating the power curve to the time when the control rod is inserted. Thus, this method is called the extrapolation method.

If the point kinetics equation with the one-precursor group model is solved, the following solution can be obtained:

$$n(t) = n_0 \left(\frac{\beta_{\mathrm{eff}}}{\beta_{\mathrm{eff}} - \rho} e^{\frac{\rho \lambda}{\beta_{\mathrm{eff}} - \rho} t} - \frac{\rho}{\beta_{\mathrm{eff}} - \rho} e^{-\frac{\beta_{\mathrm{eff}} - \rho}{\Lambda} t} \right), \tag{4.15}$$

where λ is an averaged decay constant of delayed neutron precursors in the one-precursor group approximation.

4.2.2.2 Integral Method

In the extrapolation method described above, the error induced by the extrapolation is inevitable even if measurement is precisely carried out such as setting of fine measurement point around $t = 0$. To avoid this, the integral method is also used. By adopting the Laplace transformation to Eqs. (2.54) and (2.55), the following equation is derived:

$$\overline{n}(s) = n_0 \frac{\Lambda + \sum_i \frac{\beta_{\mathrm{eff},i}}{s+\lambda_i}}{s\Lambda + s\sum_i \frac{\beta_{\mathrm{eff},i}}{s+\lambda_i} - \rho}. \tag{4.16}$$

Using

$$\lim_{s\to 0}\overline{n}(s) = \lim_{s\to 0}\int_0^\infty e^{-st}n(t)dt = \int_0^\infty n(t)dt, \tag{4.17}$$

the following equation can be derived by taking a limit of $s \to 0$ in Eq. (4.16):

$$n_0\frac{\Lambda + \sum_i \frac{\beta_{i,\text{eff}}}{\lambda_i}}{-\rho} = n_0\frac{\Lambda + \beta_{\text{eff}}\sum_i \frac{a_i}{\lambda_i}}{-\rho} = \int_0^\infty n(t)dt. \tag{4.18}$$

Generally, Λ is much smaller than $\beta_{\text{eff}}\sum\limits_i \frac{a_i}{\lambda_i}$, so the reactivity in the dollar units can be obtained from the following equation:

$$-\frac{\rho}{\beta_{\text{eff}}} = \frac{n_0\sum_i \frac{a_i}{\lambda_i}}{\int_0^\infty n(t)dt} = \frac{13.04 \times n_0}{\int_0^\infty n(t)dt}. \tag{4.19}$$

The factor 13.04 in Eq. (4.19) is derived based on the data about ^{235}U fission induced by thermal neutrons evaluated by Keepin. This is replaced by 12.75 for fission reactions with fast neutrons [2].

4.2.3 Compensation Method

In the stable period method, a positive reactivity is inserted by withdrawing a control rod (C1 here), and the reactivity is quantified from the measured stable period. In usual procedure, another control rod (C2 here) is inserted to make a reactor core in a critical state again as described in the preceding sections. The absolute value of the negative reactivity added by C2 in the second step should be equal to that of the positive reactivity added by C1. This suggests that the calibrations of C1 used to add the positive reactivity and C2 for the negative reactivity can be carried out simulataneously. The control rod calibration method for C2 in this example is known as the compensation method.

If the control rod worths of C1 and C2 are identical with each other and the positions of C1 and C2 are fully-inserted and fully-withdrawn in an initial critical state, C2 can be calibrated over an entire stroke through the calibration for C1 with the stable period method. Usually, however, worths of two control rods are different from each other, so the other methods such as the rod drop method should be also employed to terminate the C2 calibration over the entire stroke.

4.3 Experiment Procedures

4.3.1 *Reactor Core Specification*

Three control rods and three safety rods in the KUCA-C core are made of the same material and the geometric specification of them is common. An aluminum cylinder with diameter of approximately 44 mm is surrounded by a cadmium thin foil with thickness of 0.7 mm. The control rods C1, C2, and C3 are used to adjust the core reactivity while the safety rods S4, S5, and S6 are stand-by in a fully-withdrawn state during operation, and are made automatically dropped in the case of accident. Both of them can be categorized as the control rods, but their roles are different from each other.[3] Figure 4.2 shows an example of core layout at the KUCA-C core. In this layout, a set of control rods and a set of safety rods are located symmetric to each other, so the safety rods can be indirectly calibrated by the calibration of the control rods. Figure 4.3 shows detailed axial geometric specifications of the control rod and the fuel element.

In the stable period method, a positive reactivity is inserted to a critical core several times, and the sum of the inserted positive reactivity is lower than the core excess reactivity. To appropriately carry out the control rod worth calibration over an entire stroke, nuclear fuels should be added to a reactor core constructed in the approach-to-criticality experiment to increase the excess reactivity when the excess reactivity is insufficient.

Three uncompensated ionization chambers (UIC) and three fission chambers (FC) are also equipped in a core constructed through the approach-to-criticality experiment, and these are used in the control rod calibration experiment. Sometimes, auxiliary detectors such as BF_3 are also employed.

4.3.2 *Measurement with the Stable Period Method*

In the experiment with the stable period method, measurement of a doubling time T_d, which is a duration for which a value in the linear power monitor becomes double, with stopwatches is fundamental. In this measurement, T_d should be determined accurately even though there should be fluctuation. From a measured doubling time T_d, a stable period T is derived with $T = T_d/\ln 2$. In the measurement, the following should be cautioned:

(a) To confirm that the linearity of the neutron detector systems is preserved.

[3] Definition of the term "control rod" is dependent on reactor types such as critical assemblies, research reactors, and power reactors, but in the broad definitions, the term control rod is used for every neutron absorbing material located in reactor cores. In small-sized fast reactors where neutron absorbing performance is poor, nuclear fuels or neutron reflector materials are sometimes used as control rods. Control rods can be further categorized as safety rods, control rods in narrow definition, coarse adjusting rods, fine adjusting rods, etc. according to their purposes and roles.

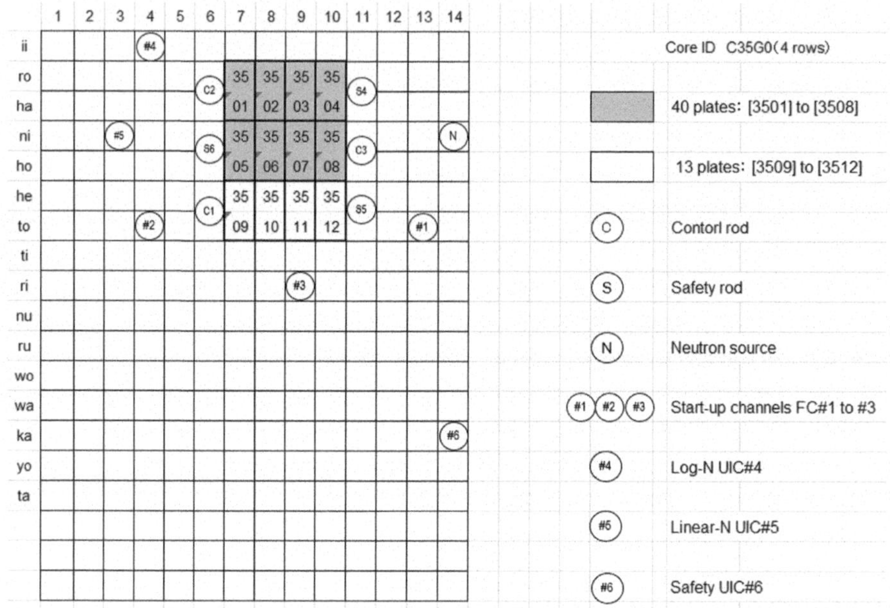

Fig. 4.2 An example of core layout in the case of highly-enriched uranium fuels

(b) To conduct the measurement with sufficiently high power level in which statistical fluctuations do not significantly affect the measurement results.

(c) To start the measurement of T_d after sufficiently long time from the reactivity insertion to avoid the contamination due to the effect of the higher-order modes, the terms from $e^{\omega_2 t}$ to $e^{\omega_7 t}$ in Eq. (2.61). Figure 4.4 shows the relative difference of the time-dependent reactor period from the stable period with several different values of T_d. This figure is prepared through numerical calculations using the physical quantities shown in Tables 2.1 and 4.1. Ragarding the prompt neutron lifetime and the effective delayed neutron fraction, the values of the C35G0 core are used. If a 5% precision is required in the measurement of the stable period, a waiting time which is longer than approximately 30 s is desired in the case of $T_d = 30$ seconds.

If the measurement is carried out in a reactor core containing nuclides causing spontaneous fission reactions such as ^{240}Pu, higher core power is required to exclude the effect of the spontaneous fission reactions. If the core power is too high, however, the temperature feedback effect causes other problems. These are not the cases at KUCA.

The following describes the procedure of the control rod calibration with the stable period method:

(1) Let us assume that the control rod C2 is a target of the calibration, and that C2 is fully inserted, C3 is fully withdrawn, and a reactor core is made in a critical

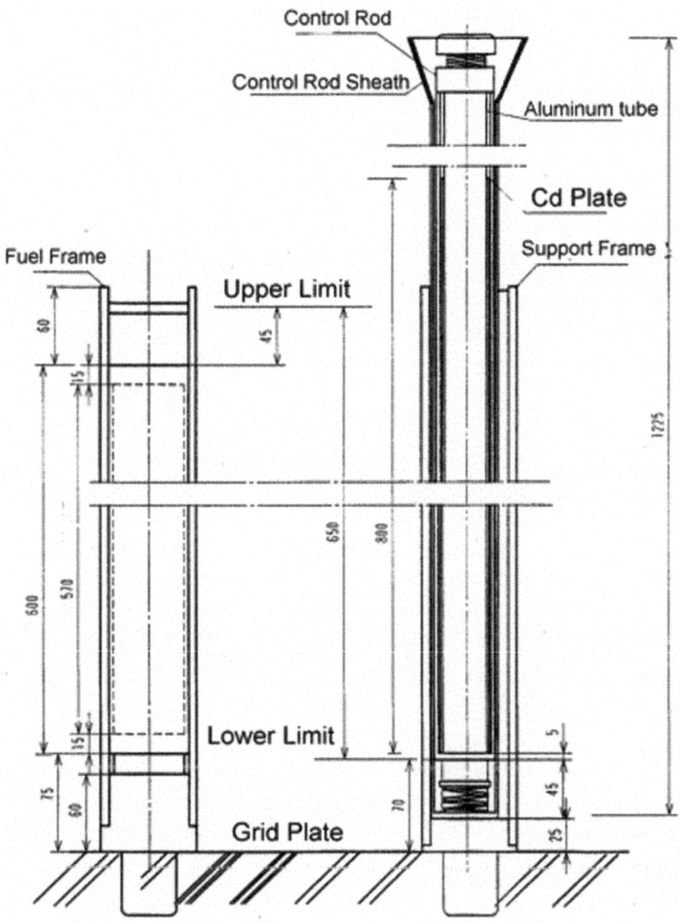

Fig. 4.3 Axial geometric specification of the control rod and the fuel plate in the KUCA-C core (unit: mm)

state with low power by adjusting the C1 position. The critical state is confirmed by observing the value in the linear power monitor on the control panel. Here it is also assumed that the core criticality is achieved in approximately 60% of the 0.01 W range of the linear power monitor.[4]

[4] Please be careful that the value indicated by the power monitor is just a rough estimation and is different from the actual power. In nuclear reactors in which core layout is easily modified such as critical assemblies, positions of neutron detectors providing the measurement data to the control panel, the UIC#5 in the KUCA-C core, can be also easily modified, so the detector efficiency is not fixed. Thus, the values in the linear power monitor are just an estimation. If it is required to obtain accurate reactor power from the linear power monitor, calibration of the linear power monitor is necessary. Please see Chap. 6.

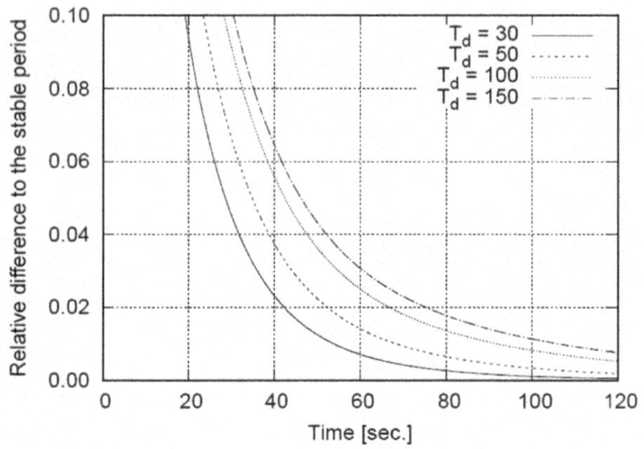

Fig. 4.4 Relative differences of the time-dependent reactor period from the stable period

(2) After confirming the criticality achievement, the range of the power range monitor is increased by two, the 0.1 W range in this case. This is to wait for sufficiently long time to observe the power increase with the stable period with high neutron count rates after the positive reactivity insertion in the next step.

(3) The C2 control rod is withdrawn to insert a positive reactivity which gives several tens to 100 s of the stable period.

(4) The reactor power begins to increase by the positive reactivity insertion, and the duration required for the power changes, 30 → 60%, 35 → 70%, and 40 → 80%, is measured with stopwatches by observing the linear power monitor with the 0.1 W range. The measured duration is the doubling time of the reactor power T_d, and thus the stable period can be easily derived from the average of the doubling time with $T = T_d/\ln 2$, and the reactivity can be determined from Eq. (4.4) or (4.9).

(5) If the reactor power reaches 110% of the present range, the control rods are automatically inserted (auto-run down). To avoid this, when the measurement of T_d is terminated below 80% of the power range, let reactor operators know that. The reactor operators at the control desk increases the range of the linear power monitor by one (0.3 W in this case) and inserts the C3 control rod to make the reactor core in a subcritical state. The reactor power should decrease.

(6) By adjusting the C3 control rod position, the reactor core becomes a critical state with low power, approximately 60% of the 0.01W range. This state is the same as the initial condition.

(7) After confirming the criticality achievement by adjusting the C3 position, C2 is withdrawn by a certain stroke to insert positive reactivity, the stable period is measured, and the reactor core is made in a critical state again by adjusting the C3 position.

(8) These operations are iterated until C2 is fully withdrawn. Doing them, C2 can be calibrated over an entire stroke with the stable period method.

At the same time, C3 is inserted from the fully-withdrawn position and the reactivity induced by this insertion is equivalent to the C2 reactivity worth. In this experiment, the reactivity of C2 measured by the stable period method is treated as the standard, and the C3 control rod is calibrated partly with the compensation method. As described above, the C3 reactivity worth is usually different from the C2 reactivity worth, so the C3 control rod cannot be calibrated over an entire stroke, and the C3 reactivity at the bottom region is still unknown in such a case. This will be concerned in the discussion (6) in Sect. 4.4.

4.3.3 Measurement with the Rod Drop Method

In this experiment, the auxiliary measurement systems such as the start-up range monitor (FC) or BF_3 and a combination of a scaler and a timer are used. Since this experiment is essentially based on the point kinetics theory, neutron detectors should be located at the spatial position where change in the neutron flux is insignificant by the control rod insertion.

(1) A reactor core should be in a critical state just after the measurement with the positive period method is terminated. As an example, let us assume that C1 is partly inserted, C2 is fully withdrawn, C3 is fully inserted, and the C2 reactivity over an entire stroke is measured with the rod drop method.
(2) Reactor power is adjusted to the level in which the deadtime[5] of the neutron measurement system is negligible: several thousand cps.
(3) After keeping a constant power for sufficiently long time, activate the scaler, and measure the neutron count rate n_0 under the present power level.
(4) By disconnecting the electric power for electrical magnets, C2 is dropped by a gravity force and is fully inserted. It is regulated at KUCA that the duration from disconnection of the electrical magnet to the initiation of rod drop (engage off) is shorter than 0.1 s and that the duration to the control rod full insertion (the seat ramp on) is shorter than 1.0 s. Generally, the rod drop is terminated for approximately 0.5 s.
(5) After waiting for sufficiently long time and confirming that the reactor power reaches the background level, the measurement is terminated.
(6) The same procedure is adopted to the other control rods to calibrate each rod.

[5] The dead time is the minimum time duration for which two subsequent events can be detected as two independent pulses by a neutron detector. Events such as fission reactions occurring in a detector during the dead time are not detected. The dead time is caused by both of the detection process of the detector itself and the electric circuits connecting with the detectors. The behavior of the dead time can be classified into two types of the paralyzable behavior and the non-paralyzable behavior. In the latter, true count rate n_0 can be estimated from $n_0 = \frac{n}{1-n\tau}$ where n is measured count rate and τ is the deadtime, and thus the contribution of the neutron detection during the dead time can be corrected.

In the case of the extrapolation method, n_1 is determined through the extrapolation based on the measurement results, and the reactivity is derived with Eq. (4.14).

In the integral method, the neutron count rate n_0 is measured with the scaler with the appropriate measurement time at first. Then the neutron count measurement is started with the scaler with the free preset time at the same time when the electric power is disconnected with the electrical magnet. The neutron count should decrease with time, so the neutron count measurement is finished when the neutron count rate reaches the background level. The measurement result obtained with this step is used as the denominator of Eq. (4.19). Note that the correction of the background does not affect the final results since these measurements are started under relatively high-neutron flux level conditions.

4.4 Discussions

Items for the exercise are as follows:

(1) In the measurement of the stable period method, derive the reactivity ρ by hands from the stable period T measured with stopwatches and Eq. (4.3), and summarize the results with the table in which the contributions of the group-wise delayed neutrons to the reactivity are clearly described. Please ask instructors which measurement is concerned.
(2) Obtain the control rod reactivity worth from the measurement results with the stable period method and the rod drop method (the extrapolation method and the integral method).
(3) Obtain the integral and differential reactivity curves for the control rod to which the measurement is carried out with the stable period method.
(4) Derive the shutdown margin.
(5) Derive the excess reactivity.
(6) Derive the reactivity worth per one fuel plate from the result in the exercise (5) and the excess reactivity of the reactor core in the approach-to-criticality experiment.

The items for discussions are as follows:

(1) Based on the results of the exercise (2), discuss the difference in the reactivity worth among different control rods qualitatively and quantitatively. The control rod reactivity worth derived from the assumption of the cosine-shaped neutron flux distribution discussed in Sect. 4.6 would be useful if it is applied to the radial x–y plane of the core.
(2) Compare the reactivity worth deduced from the stable period method with that from the rod drop method. If there is a difference, discuss its reason.
(3) Compare the reactivity worth deduced from the integral method with that from the extrapolation method. If there is a difference, discuss its reason.

(4) Let $\rho(H)$ be the control rod worth deduced from the stable period method and plot a semi-analytical control rod (integral) worth curve using Eq. (4.20) in Sect. 4.5 with the measurement-based results obtained in the exercise (3). Please be careful that the measurement results are on the "withdrawal length" and the results with Eq. (4.20) are on the "insertion length".

(5) Compare the two curves obtained in the item (4) with each other, and if there is a difference, discuss the reason. Please be careful on the following; Eq. (4.20) is derived for a bare reactor in which neutron flux spatial distribution is approximated by the sine/cosine function, so the parameters H and x should be carefully examined. Results obtained with the experiment for the neutron flux spatial distribution would be also useful. The axial geometric configurations of the control rod and the fuel plate in the KUCA-C core can be found in Fig. 4.3.

(6) Let us assume that the C2 control rod is calibrated by the stable period method and the C3 control rod is used for the reactivity compensation. When the C2 reactivity worth is slightly larger than the C3 reactivity worth, the reactor core does not reach a critical state when C2 is fully withdrawn and C3 is fully inserted. Describe how to terminate the C3 calibration over an entire stroke in such a case.

(7) Control rod calibration is generally performed with the positive stable period, and the negative stable period is generally not used. Discuss the reason.

4.5 Preliminary Exercises

Before the experiment, it is desired that the following exercises are tackled.

(1) As shown in Fig. 4.5, let us assume that C2 is fully inserted, C3 is fully withdrawn, and the reactor core is in a critical state by adjusting the C1 position. C2 and C3 are calibrated with the stable period method and the compensation method, respectively. When we assume to calibrate C2 with three withdrawal actions, prepare a conceptual figure showing the measurement procedure with control rod movement as Fig. 4.5. The C2 and C3 worth are assumed identical with each other.

(2) Based on the one-group first-order perturbation theory, reactivity induced by control rod insertion of a length x from the top position can be represented as follows [1]:

$$\rho(x) = \rho(H)\left(\frac{x}{H} - \frac{1}{2\pi}\sin\left(\frac{2\pi x}{H}\right)\right), \tag{4.20}$$

where $\rho(H)$ is a reactivity induced by the full insertion from the top position. The derivation of this equation is described in Sect. 4.6. Plot $\rho(x)$ using Eq. (4.20) with H=650.0 and $\rho(H)$=1. Note that the full stroke length of the control rod drive mechanism of the KUCA-C core is 650.0 mm. Summarize the assumptions introduced to derive Eq. (4.20) also.

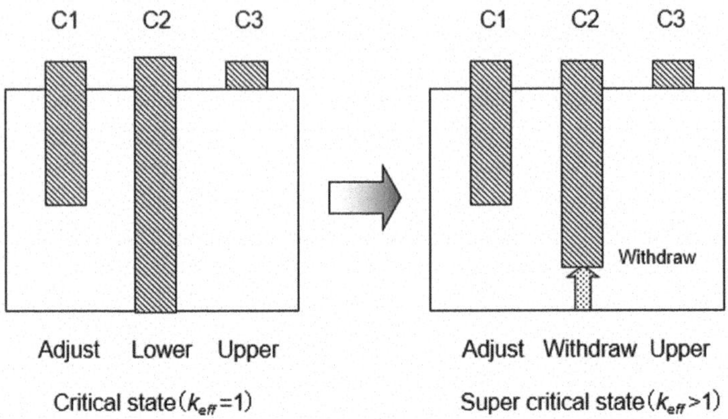

Fig. 4.5 Conceptual figure on control rod insertion position and control rod manipulation (Side view of reactor core with three control rods, C1, C2, and C3)

(3) The step (3) in Sect. 4.3.2 describes that the C2 control rod is withdrawn to insert a positive reactivity with a stable period of several tens to 100 s. The length of the control rod stroke satisfying this condition is dependent on a control rod insertion position and is not constant over an entire stroke. Based on the results of the above item (2), estimate the stroke length per one control rod withdrawal assuming the control rod reactivity worth, $\rho(H)$=0.4 × 10^{-2} $\Delta k/k$. Also, determine how many withdrawal actions are required to calibrate this control rod.

(4) A stable period T after a reactivity insertion is given in Eq. (4.2), and this is also applicable to the case of the negative large reactivity insertion in the rod drop method. In such a case, since ω_1 becomes close to $-\lambda_1$, the period $T = 1/\omega_1$ becomes $-1/\lambda_1$ ~$-$80 s [1]. In the rod drop method with the integral method, neutron count measurement is carried out for a finite measurement time after the control rod insertion, and this means the range of the time integral in the denominator of Eq. (4.19) is limited. Calculate and compare the integrated neutron counts from 300 to 600 s, from 600 to 900 s, and from 900 to infinite seconds using $T \approx -80$ seconds, and discuss the error induced by the finite measurement time.

4.6 The First-Order Perturbation Theory and Control Rod Calibration Curve

When we calculate a change in the effective multiplication factor k_{eff} (simply k in the following) induced by a slight change in a core size and/or composition, the perturbation method can be applied instead of the criticality calculations with the iterative methods if the change is sufficiently small.

First, let us consider the following equation governing neutron transport and multiplication:

$$M\phi = \frac{1}{k}F\phi, \tag{4.21}$$

where M is the neutron disappearance/scattering operator and F is the neutron generation operator. In the case of the one-group diffusion equation, these operators are represented as

$$M = -\nabla D(\mathbf{r})\nabla + \Sigma_{\mathrm{a}}(\mathbf{r}),\ F = \nu\Sigma_{\mathrm{f}}(\mathbf{r}). \tag{4.22}$$

Let us consider that a neutron absorber is locally introduced to this reactor and the macroscopic absorption cross section is changed to Σ'_{a} as

$$\Sigma'_{\mathrm{a}}(\mathbf{r}) = \Sigma_{\mathrm{a}}(\mathbf{r}) + \delta\Sigma_{\mathrm{a}}(\mathbf{r}). \tag{4.23}$$

Due to this, neutron flux ϕ and the effective multiplication factor k should be changed, and the following another equation is satisfied:

$$M'\phi' = \frac{1}{k'}F\phi', \tag{4.24}$$

$$M' = M + \delta\Sigma_{\mathrm{a}}. \tag{4.25}$$

Let us introduce the adjoint neutron flux ϕ^+ and the adjoint operators M^+ and F^+. These satisfy the following equations:

$$M^+\phi^+ = \frac{1}{k}F^+\phi^+, \tag{4.26}$$

$$\langle M^+f, g\rangle = \langle f, Mg\rangle, \tag{4.27}$$

where the brackets < > are for the integration over an entire volume. By calculating the inner product between each term in Eq. (4.24) and ϕ^+ and using Eq. (4.25), the following equation can be derived:

$$\langle\phi^+, M\phi'\rangle + \langle\phi^+, \delta\Sigma_a\phi'\rangle = \frac{1}{k'}\langle\phi^+, F\phi'\rangle. \tag{4.28}$$

Using the equation defining the adjoint operators (4.27), the first term in the left-hand side of Eq. (4.28) can be transformed into

$$\langle\phi^+, M\phi'\rangle = \langle M^+\phi^+, \phi'\rangle = \frac{1}{k}\langle F^+\phi^+, \phi'\rangle = \frac{1}{k}\langle\phi^+, F\phi'.\rangle \tag{4.29}$$

By subtracting Eq. (4.28) from Eq. (4.29), the following equation for the reactivity change $\Delta\rho$ can be derived:

$$\Delta\rho = \frac{1}{k} - \frac{1}{k'} = -\frac{\langle \phi^+, \delta\Sigma_a \phi' \rangle}{\langle \phi^+, F\phi' \rangle}. \tag{4.30}$$

No approximations have been introduced to derive Eq. (4.30). If we assume that the changes in both the neutron flux and the cross sections are small, $\delta\Sigma_a\phi' \approx \delta\Sigma_a\phi$ and $\langle \phi^+, F\phi' \rangle \approx \langle \phi^+, F\phi \rangle$, the reactivity change $\Delta\rho$ can be obtained as

$$\Delta\rho = -\frac{\langle \phi^+, \delta\Sigma_a \phi \rangle}{\langle \phi^+, F\phi \rangle}. \tag{4.31}$$

This equation includes only the neutron flux, the adjoint flux, and the operators before the cross section change and does not require information after the cross section change. This is known as the first-order perturbation approximation.

Furthermore, in the case of the one-group approximation, the adjoint operators such as M^+ are identical to the normal operators and the neutron flux is also identical to the adjoint neutron flux.

Thus, Eq. (4.31) can be explicitly represented as

$$\Delta\rho = -\frac{\int_V \phi(\mathbf{r})\delta\Sigma_a(\mathbf{r})\phi(\mathbf{r})dV}{\int_V \phi(\mathbf{r})\nu\Sigma_f(\mathbf{r})\phi(\mathbf{r})dV}. \tag{4.32}$$

The numerator in this equation is an integral over regions where the macroscopic absorption cross sections are changed. When absorbers are added to a homogeneous volume ΔV, the reactivity is calculated as

$$\Delta\rho = -\frac{\delta\Sigma_a}{\nu\Sigma_f \int_V (\phi(\mathbf{r}))^2 dV} \int_{\Delta V} (\phi(\mathbf{r}))^2 dV. \tag{4.33}$$

The above first-order perturbation theory is often used to calculate reactivity effect induced by introduction of materials into a reactor core. As an example, let us consider a reactivity calibration curve of a control rod. For simplicity, a homogeneous (constant cross section over an entire volume) infinite slab reactor core with a thickness of H without any neutron reflectors is considered, and reactivity induced by the control rod insertion in the spatial position of $0 \leq x \leq H$ is calculated by the one-group first-order perturbation theory. In this model, the neutron flux spatial distribution can be represented as

$$\phi(x) = A \sin\frac{\pi x}{H}, \tag{4.34}$$

where A is a constant. As shown in this equation, neutron fluxes at $x = 0$ and $x = H$ are zero. Small change in the macroscopic absorption cross section in Eq. (4.25) is represented as

$$\delta\Sigma_a(x) = \begin{cases} \delta\Sigma_a(0 \le x \le h) \\ 0(h < x \le H) \end{cases}. \tag{4.35}$$

By substituting Eqs. (4.34) and (4.35) into Eq. (4.32), the following equation is derived:

$$\Delta\rho(h) = -\frac{\delta\Sigma_a \int_0^h \sin^2 \frac{\pi x}{H} dx}{\nu\Sigma_f \int_0^H \sin^2 \frac{\pi x}{H} dx} = -\frac{\delta\Sigma_a}{\nu\Sigma_f}\left(\frac{h}{H} - \frac{1}{2\pi}\sin\frac{2\pi h}{H}\right). \tag{4.36}$$

In the case of $h = H$ in Eq. (4.36), the reactivity induced by full insertion of the control rod $\Delta\rho(H)$ can be derived as

$$\Delta\rho(H) = -\frac{\delta\Sigma_a}{\nu\Sigma_f}, \tag{4.37}$$

and thus, according to Eqs. (4.36) and (4.37), the reactivity induced by the control rod insertion over $0 \le x \le H$ can be written as

$$\Delta\rho(h) = \Delta\rho(H)\left(\frac{h}{H} - \frac{1}{2\pi}\sin\frac{2\pi h}{H}\right). \tag{4.38}$$

This is a reactivity calibration curve in the one-dimensional simple slab reactor core model.

References

1. Lamarsh JR (1966) Introduction to nuclear reactor theory. Addison-Wesley Publishing Company, Reading, Massachusetts
2. Keepin GR (1965) Physics of nuclear kinetics. Addison-Wesley Publishing Company, Reading, Massachusetts

Chapter 5
Measurements of Reaction Rates

Abstract Neutron flux plays a significant role in the theory of reactor physics, while neutron flux is indirectly measured through the interaction between the neutron and the material used as the neutron detector in experiments. The physical quantity that is measured throughout the experiment is the "reaction rate." To obtain the neutron flux from the measured reaction rate, various information involving cross sections of the material used as the detector is necessary, in addition to experimental devices. At the same time, it is necessary to consider the theory of reactor physics related to the specific reactions between neutrons and materials in the irradiation field. This chapter introduces methodologies for measuring the reaction rate by the activation foil method before determining the neutron flux.

Keywords Reaction rate · Activation foil method · Saturated activity · Gold wire · Cadmium ratio

5.1 Overview

5.1.1 Background

The spatial distribution of neutron flux in the core can be uniquely determined by the type and structure of a reactor, the composition and shape of nuclear fuel, and the locations of control rods, when a nuclear reactor is maintained in a critical state and operated at a constant power. In light-water reactors, which are typical thermal reactors, most fission reactions are caused by thermal neutrons, and the spatial distribution of fission reaction rates can be considered almost proportional to the spatial distribution of the thermal neutron flux. Furthermore, since the power distribution of a reactor is proportional to the fission reaction rate distribution, the thermal neutron flux distribution and the power distribution are almost proportional to each other. If the relative spatial distribution of thermal neutron flux and the absolute value of thermal neutron flux are measured at appropriate regions and locations, respectively, these two can be combined to obtain the absolute value of the thermal reactor power.

C. H. Pyeon et al., *Reactor Laboratory Experiments at Kyoto University Critical Assembly*, https://doi.org/10.1007/978-981-97-8070-9_5

Thermal neutron flux can be measured by using activation detectors with high sensitivity to thermal neutrons. To measure high accuracy of thermal neutron flux, it is important to be unaffected by the high γ-ray dose in the core and to have high accuracy and spatial resolution of measurements with respect to positions. Considering the characteristics of detection efficiency and spatial resolution, it is said that the activation detector is the most appropriate to measure thermal neutron flux without disturbing the spatial distribution of neutron flux.

5.1.2 *Objectives*

The gold-197 (^{197}Au), which has high sensitivity to thermal neutrons, is often used to measure reaction rates in the core. In the light-water-moderated and light-water-reflected core (C core) at KUCA, reaction rate distributions in the core are generally obtained through the capture reactions between ^{197}Au and neutrons.

When learning about measurements of reaction rates introduced in this chapter, if you keep the following objectives in mind as you read, you will gain a deeper understanding of measurements of reaction rates, involving the relationship between the reaction rate and neutron flux, and will also recognize the importance of reaction rate measurements. The main objectives of experiments are shown as follows:

- To measure the reaction rate distribution corresponding to the neutron flux distribution in the core using the activation of Au wire.
- To acquire basic knowledge of activation foil method through experiments.
- To discuss the effects of control rods, safety rods, light-water reflector, and component materials in the core on reaction rate distributions.
- To gain insight into the neutron spectrum in the core, resulting in the reaction rates by the bare Au wire, and the Au wire inserted into a cadmium (Cd) tube (Cd-tubed Au wire).

5.2 Activation Detectors

5.2.1 *Overview*

Neutrons can be measured indirectly by any detectors, as mentioned in the abstract of this chapter. The basic principle of measuring neutrons is to obtain information on neutrons indirectly through the measurement of radiation generated by the interaction between the neutron and a certain material. When a sample of a certain material is irradiated by neutrons in a reactor or accelerator for a certain period, information on the number and energy distribution of neutrons in the irradiation field can be extracted by measuring the induced radioactivity of the radiation generated there. The sample of material used in this process is called an activation detector.

Activation detectors are most widely used to measure thermal neutrons, because reaction cross sections between a material and a neutron are large in the low-energy region (mainly on the order of eV) owing to the $1/\upsilon$ law (Fig. 6.4 in Chap. 6). To obtain high sensitivity to thermal neutrons, it is necessary to select a material with large reaction cross sections. It is, however, desirable to reduce the thickness and geometry of the material to avoid disturbing the neutron flux field to be measured and to minimize the self-shielding effect (see Sect. 6.4.4), since the mean free path of neutrons in a material with large reaction cross sections is very small. Mainly for the reasons given above, a thin foil or fine wire is chosen as the sample material used for the measurement.

5.2.2 *Characteristics*

The activation detector is used for obtaining reaction rates, whereas time-series data of neutron flux cannot be obtained from an irradiated foil or wire. Due to its geometrical shape and small thickness, the following advantages of activation detectors can be mentioned:

- Sample is small in dimensions and inexpensive.
- Foil mass and dimensions are smaller than those of a typical neutron detector, so the foil or wire is less likely to disturb the neutron irradiation field to be measured, and furthermore, it has superior spatial resolution.
- Neutron irradiation is possible even in extreme environments such as a high-intensity γ-ray field are where neutron detectors cannot be used, and electrical contact with the outside world is not required.
- Accuracy of neutron flux measurements is high, when the interaction of neutrons with a certain of material is known.
- Neutron signals are measured without the influence of nuclear reactions by the background γ-ray influence.[1]
- The equipment for measuring induced radioactivity is simple and the cost of the equipment is low.
- The activation foil has a wide range of detection sensitivity and is relatively easy to deal with regardless of the intensity of the neutron flux.[2]

[1] Only when nuclear reactions by γ-ray in the material are negligible compared to nuclear reactions by neutrons.

[2] The irradiation time should be adjusted so that the radioactivity of radioisotope produced by activation reactions can be measured properly. Usually, if the value of neutron flux is large, the irradiation time of activation foil should be shortened, and if it is small, the irradiation time should be lengthened. However, if the irradiation time is too short, it is necessary to take a certain amount of time for obtaining high power in a reactor operation. Therefore, the ratio of radioisotopes produced during the rise or fall of the reactor power is negligible. On the other hand, if the irradiation time is too long, the fluctuation of reactor power during irradiation easily affects.

On the other hand, compared with general neutron detectors, the following disadvantages can be mentioned:

- The energy distribution of neutrons is not obtained without using more than two kinds of activation detectors with threshold reactions.[3]
- Data processing is complicated due to the need for various corrections and calibrations.
- It takes a considerable amount of time from the start of irradiation until the results are obtained, and it is difficult to conduct on-line measurements.
- The use of nuclides for which detailed data on activation cross sections are not available will inevitably cause the low accuracy of the measurement.

5.2.3 Physical Properties

In selecting materials for activation detectors, nuclear and physical properties should be fully considered as follows:

(1) Energy dependence of cross sections

In general, an activation detector (foil and wire) consists of some elements. One of these elements should have large capture cross sections to thermal neutrons (low-energy neutrons), resulting in radioisotope after neutron irradiation. In addition, most reactions with thermal neutrons are dominated by capture reactions because capture cross sections follow approximately the $1/\sqrt{E}$ law (or $1/\upsilon$ with respect to velocity υ) for energy E in the low-energy region. However, many nuclei have resonance peaks of capture cross sections mainly at certain neutron energies between 1 to 1000 eV. Therefore, when a nucleus is activated by thermal neutrons, it may also contain isotopes activated by neutrons in the resonance energy region.

(2) Half-life of radioisotopes

Radioisotope produced by neutron irradiation emits radiation, and the intensity of radioactivity should be high to have a measurement accuracy. Since the intensity of radioactivity is obtained by the product $N\lambda$ of the number of the radioisotope in the activation detector N and its decay constant λ ($\lambda = \ln 2/T_{1/2}$, where $T_{1/2}$ is the half-life), the half-life of the radioisotope is one of the important factors in determining the irradiation time by neutrons. For example, if the half-life is short, the intensity of radioactivity is high, but it decays quickly, so it is often difficult to measure the radioisotope after irradiation.

(3) Isotope abundance and purity

Activation detectors consist of many elements: we can categorize these as an isotope to be activated by neutron capture and impurities. The isotope concentration in the

[3] Cross sections of activation reactions continuously vary in response to neutron energy, and some nuclides show sharp peaks, such as resonance reactions, or have threshold reactions. There is, however, no such thing as a selective reaction for neutrons with a certain energy.

activation detector, in general, is not high: concentrating an isotope costs a lot. The activation detector is then desirable to have a high isotope abundance ratio in the natural of the activation detector. In addition, when impurities in the sample are activated, it is difficult to measure the radiation, and the impurity that causes the problem should be removed. Therefore, it is required that samples of high purity should be used in the activation detector.

(4) Properties of emitted radiation

The decay of nuclei produced in activation reactions involves the emission of β-ray or γ-ray. The γ-ray is said to be suitable for measurement because its permeability is strong and self-absorption effects are small in the sample. Furthermore, measuring the energy spectrum of γ-ray is easier than that of β-ray.

From these properties shown, Table 5.1 summarizes the various properties of activation materials generally used for measurements of thermal neutrons with respect to capture reactions.

Table 5.1 Activation materials for thermal neutron detection and their various properties for capture reactions (Refs. [1, 2])

Isotope (abundance ratio: %)	Radioisotope	Half-life	γ-ray energy (MeV) (emission rate: %)	Cross section* (barn)**
^{55}Mn (100)	^{56}Mn	2.5789 h	0.846 (99)	13.4 ± 0.1
^{59}Co (100)	^{60}Co	5.2714 y	1.173 (100)	20.2 ± 1.9
			1.333 (100)	
^{63}Cu (69.1)	^{64}Cu	12.700 h	0.511	4.41 ± 0.20
^{65}Cu (30.9)	^{66}Cu	5.088 m	1.039	1.8 ± 0.4
^{107}Ag (51.35)	^{108}Ag	2.37 m	0.633	45 ± 4
^{109}Ag (48.35)	^{110m}Ag	249.75 d	0.658 (94.0)	3.2 ± 0.4
			0.884 (72.19)	
^{115}In (95.77)	^{116m}In	54.41 m	0.417 (32)	160 ± 2
			1.097 (56)	
			1.293(84)	
^{164}Dy (28.18)	^{165m}Dy	1.257 m	0.108	2651 ± 265
			0.515	
	^{165}Dy	2.334 h	0.094 (3.6)	800 ± 100
			0.361	
^{197}Au (100)	^{198}Au	2.69517 d	0.411 (96)	98.6 ± 0.4

*For neutron velocity 2200 m s^{-1} (thermal neutrons: neutron energy 0.0253 eV)
**1 barn = 10^{-24} cm^2

5.3 Measurements of Reaction Rates

Assuming that a substance of mass M exists in the core, and that this substance consists of isotope a of atomic weight A with existence ratio W, and using the Avogadro number N_0 (mol^{-1}), the number of atoms of isotope a in the substance, N_a, can be obtained as follows:

$$N_a = \left(\frac{M\,W}{A}\right) N_0. \tag{5.1}$$

We consider the use of isotope a as an activation foil, taking advantage of the reaction in which this isotope a is activated by neutrons in the reactor core to be the radioisotope b.

5.3.1 *Activation Reaction Rates*

For neutrons in the energy range between E and $E + dE$ (MeV), the microscopic activation cross section of isotope a is $\sigma_a(E)$ (cm^2) and the neutron flux at the location of the activation foil is $\phi(E)dE$ (cm^{-2} s^{-1}). The neutron flux in the activation foil is assumed to be undistorted. The number of reactions in which isotope a is activated to radioisotope b per unit time, i.e., the activation reaction rate $R(E)dE$ (s^{-1}), can be expressed using the number of atoms N_a as follows:

$$R(E)dE = N_a \sigma_a(E)\phi(E)dE. \tag{5.2}$$

If the maximum energy of neutrons in the core is $E_{\max}$ (MeV), the total activation reaction rate of the activation foil R_{act} (s^{-1}) can be expressed using Eq. (5.2) as follows:

$$R_{act} = \int_0^{E_{\max}} R(E)dE = \int_0^{E_{\max}} N_a \sigma_a(E)\phi(E)dE. \tag{5.3}$$

At the location where the activation foil is set, the total neutron flux Φ (cm^{-2} s^{-1}) and the effective microscopic activation cross section of the activation foil $\sigma_{\mathrm{eff},a}$ (cm^2) are defined as follows, respectively:

$$\Phi = \int_0^{E_{\max}} \phi(E)dE, \tag{5.4}$$

$$\sigma_{\text{eff},a} = \int_{0}^{E_{\max}} \frac{\sigma_a(E)\phi(E)dE}{\Phi}. \tag{5.5}$$

Using Eqs. (5.4) and (5.5), Eq. (5.3) can be expressed as follows:

$$R_{act} = N_a \sigma_{\text{eff},a} \Phi. \tag{5.6}$$

Aside from the absolute value of the neutron spectrum $\phi(E)$ ($\text{cm}^{-2}\,\text{s}^{-1}\,\text{MeV}^{-1}$), if the spatial distribution of $\phi(E)$ is known, $N_a \sigma_{\text{eff},a}$ in Eq. (5.6) is specific to the activation foil. The reaction rate R_{act} can be then quantified by measuring the radioactivity (decay rate) of the radioisotope b, resulting in the total neutron flux Φ. This is the principle of neutron flux measurement by activation foil.

5.3.2 *Saturated Activity*

As is well known, in the activation foil, not only the production of the radioisotope b occurs, but also the radioactive decay is taken place. If the decay constant of radioisotope b is λ_b (s^{-1}) and the number of atoms at time t (s) is $N_b(t)$, the number of b decaying per unit time is $\lambda_b N_b(t)$. Also, due to production and decay of radioisotope b occurring in the activation foil between time range t and $t + dt$ (s), the change in the number of atoms can be expressed as follows:

$$dN_b(t) = N_a \sigma_{\text{eff},a} \Phi dt - \lambda_b N_b(t) dt. \tag{5.7}$$

Since there is no radioisotope b before irradiation starts, using the initial condition $N_b(0) = 0$ at $t = 0$, the number of atoms of radioisotope $N_b(T_i)$ at the irradiation time T_i (s) can be obtained from Eq. (5.7) as follows:

$$N_b(T_i) = \frac{N_a \sigma_{\text{eff},a} \Phi \left(1 - e^{-\lambda_b T_i}\right)}{\lambda_b}. \tag{5.8}$$

Using Eq. (5.8) and the decay constant λ_b, the induced radioactivity $D_b(T_i)$ (s^{-1}) of radioisotope b after the irradiation time T_i can be expressed as follows (time correction for irradiation time):

$$D_b(T_i) = \lambda_b N_b(T_i) = N_a \sigma_{\text{eff},a} \Phi \left(1 - e^{-\lambda_b T_i}\right). \tag{5.9}$$

The radioactivity of radioisotope b produced when irradiated for infinite time, i.e., the saturated activity D_∞ (s^{-1}), can be expressed by applying $T_i \to \infty$ to Eq. (5.9) as follows:

$$D_\infty = N_a \sigma_{\text{eff},a} \Phi. \tag{5.10}$$

The relationship between Eqs. (5.6) and (5.10) is then established as follows:

$$D_\infty = R_{act}. \tag{5.11}$$

This indicates that the saturated activity is equal to the total activation reaction rate, i.e., to find the total activation reaction rate R_{act}, the saturated activity D_∞ should be measured.

5.3.3 *Time and Weight Corrections*

The number of radioisotope atoms produced by irradiation decreases with time after the end of irradiation, and the radioactivity $D_b(T_i)$ at that time also decreases with time. If the waiting time is T_w (s) from the end of irradiation to the start of radioactivity measurement, the radioactivity $D_b(T_w)$ (s^{-1}), when the measurement of the γ-ray spectrum is started, can be expressed as follows (time correction regarding waiting time to start measurement):

$$D_b(T_w) = D_b(T_i) e^{-\lambda_b T_w}. \tag{5.12}$$

Using the radioactivity $D_b(T_w)$ at the start of the measurement, the detection efficiency ε_D of the radiation detector (e.g. high-purity germanium: HPGe detector or sodium iodide (thallium): NaI(Tl) detector) and the emission rate of the γ-ray energy ε_E (see Table 5.1), the number of counts c_{total} of the γ-ray spectrum in the measurement time T_c (s) can be obtained as follows:

$$c_{total} = \int_0^{T_c} \varepsilon_\text{D} \varepsilon_\text{E} D_b(T_w) e^{-\lambda_b t} dt = \frac{\varepsilon_\text{D} \varepsilon_\text{E} D_b(T_w)\left(1 - e^{-\lambda_b T_c}\right)}{\lambda_b}. \tag{5.13}$$

Using Eq. (5.13), the average count rate C_{av} (s^{-1}) at the measurement time T_c can be expressed as follows (time correction for measurement time):

$$C_{av} = \frac{c_{total}}{T_c} = \frac{\varepsilon_\text{D} \varepsilon_\text{E} D_b(T_w)\left(1 - e^{-\lambda_b T_c}\right)}{\lambda_b T_c}. \tag{5.14}$$

As shown above, after making corrections for the irradiation time T_i (s), the waiting time T_w (s) and the measurement time T_c (s) as shown in Eqs. (5.9), (5.12) and (5.14), respectively, the saturated activity D_∞ (s^{-1}) can be expressed as follows:

$$D_\infty = \frac{\lambda_b T_c C_{av}}{\varepsilon_D \varepsilon_E \left(1 - e^{-\lambda_b T_i}\right) e^{-\lambda_b T_w} \left(1 - e^{-\lambda_b T_c}\right)}. \tag{5.15}$$

Here, we assume that the total activation reaction rate per volume $R_{reaction}$ (cm^{-3} s^{-1}), in the strict sense, is equal to the saturated activity. For the activation foil, however, it is necessary to make a correction for the weight of the activation detector. When the weight and density of the activation detector are M (g) and ρ (g cm^{-3}), respectively, the total activation reaction rate per volume $R_{reaction}$ (cm^{-3} s^{-1}) can be expressed using the saturated activity D_∞ (s^{-1}) in Eq. (5.15) as follows (weight correction with respect to sample):

$$R_{reaction} = D_\infty \frac{\rho}{M} = \frac{\lambda_b T_c C_{av} \rho}{\varepsilon_D \varepsilon_E M \left(1 - e^{-\lambda_b T_i}\right) e^{-\lambda_b T_w} \left(1 - e^{-\lambda_b T_c}\right)}. \tag{5.16}$$

When determining the relative spatial distribution of the total activation reaction rate, it is not necessary to obtain the absolute values of D_∞ of all samples. Namly, to obtain the relative spatial distribution, it is not necessary to know the detection efficiency ε_D and the emission rate ε_E of γ-ray energy shown in Eq. (5.16), and the density of the activation foil ρ. In addition, since the decay constant λ_b and the irradiation time T_i are the same for the activation foils to be measured, it is only necessary to consider the corrections related to the waiting time T_w and the measurement time T_c. Based on the above, the relative value of the total activation reaction rate can be approximated using Eq. (5.16) as follows:

$$R_{reaction} \approx \frac{T_c C_{av}}{M e^{-\lambda_b T_w} \left(1 - e^{-\lambda_b T_c}\right)}. \tag{5.17}$$

From the results of the derivation of Eq. (5.17), the relative reaction rate by the activation detector can be determined.

5.4 Cadmium Ratio

5.4.1 Cadmium Difference Method

When activation foils are irradiated in a thermal neutron reactor, the activation reaction rate is inevitably affected not only by thermal neutrons but also by epi-thermal neutrons, so it is necessary to remove the effect of epi-thermal neutrons to extract information on thermal neutrons only. The Cd difference method is widely used in thermal neutron reactors that makes good use of the unique energy dependence of Cd capture cross sections.

The cadmium has large capture cross sections for low-energy neutrons and a large resonance absorption (peak) around 0.18 eV, as shown in Fig. 5.1. Note that

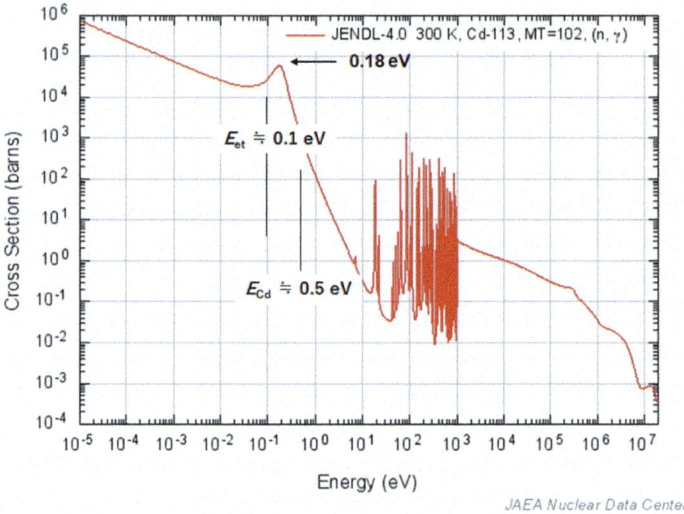

Fig. 5.1 Capture cross sections of ^{113}Cd (data from JENDL-4.0 in Ref. [3])

the resonance energy is relatively close to the boundary energy between the thermal and epi-thermal regions, i.e., the thermal spectrum cut-off energy $E_{et} \approx 0.1$ eV, and that capture cross section is very large in the energy region corresponding to the Maxwellian distribution.

Using the above, by inserting the activation wire into a fine Cd tube of appropriate thickness (hereinafter termed Cd tube),[4] all neutrons less than a certain energy E_{Cd} (hereinafter termed effective Cd cut-off energy) can be absorbed, and activation by neutrons less than E_{Cd} can be prevented. In other words, only neutrons with energies greater than E_{Cd} will be penetrated into the activation wire, and only activation by neutrons greater than E_{Cd} will be considered to occur.

Since activation by neutrons of all energies occurs in the activation wire without the Cd tube (hereafter termed bare), we can subtract the activation reaction rate of the Cd-tubed activation wire from that of the bare wire, to obtain the activation reaction rate due only to neutrons of energy less than E_{Cd}. Furthermore, by adding corrections due to the difference between E_{Cd} and the thermal spectrum cut-off energy E_{et}, and corrections for the contribution of thermal neutrons transmitted through the Cd tube, as described in Sect. 5.4.2, the activation reaction rate induced by thermal neutrons only can be obtained. In other words, the principle of the Cd difference method is to obtain the activation reaction rate obtained by thermal neutrons only, finding the difference between the bare and Cd-tubed activation wires and applying the appropriate corrections.

[4] For the Au foil, the Cd cover is used.

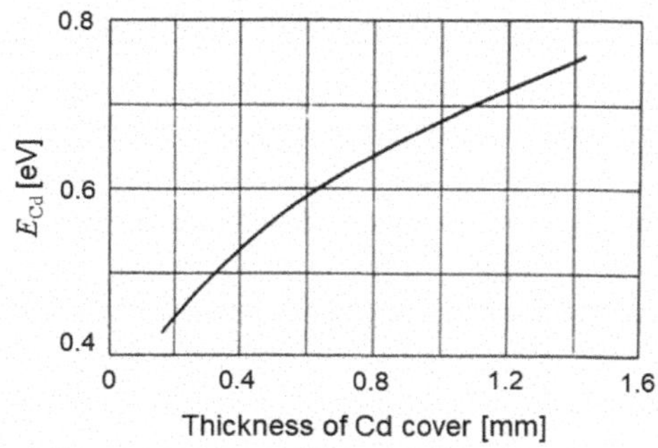

Fig. 5.2 Variation of effective Cd cut-off energy E_{Cd} with Cd cover thickness for $1/\upsilon$-type absorber (Ref. [4])

5.4.2 Effective Cadmium Cut-Off Energy

In the Cd difference method, the activation reaction rate obtained by thermal neutrons is determined based on the difference in the activation reaction rate bordered by the effective Cd cut-off energy E_{Cd}. This effective Cd cut-off energy E_{Cd} is defined below, and noteworthy is that E_{Cd} is dependent on the thickness of Cd cover (tube).

Assume a step approximation that Cd absorption cross sections $\sigma_c(E)$ (microscopic capture cross sections; see Fig. 6.3 in Chap. 6) is infinity at $E \leq E_{Cd}$ and zero at $E > E_{Cd}$. The E_{Cd} is then determined by assuming that the activation reaction rate obtained by the neutron spectrum with the $1/E$-type distribution (see Chap. 6) in the region $E_{Cd} < E < \infty$ is equal to that in the full energy region $0 \leq E < \infty$. Here, calculation results of E_{Cd} for a very thin $1/\upsilon$-type activation foil are shown in Fig. 5.2.

5.4.3 Cadmium Ratio

When the activation wire inserted into Cd tube is irradiated, the activation reaction is considered to be caused almost exclusively by epi-thermal neutrons of energy E ($> E_{Cd}$). For the activation reaction rates R_{bare} and R_{Cd} of the bare and Cd-tubed Au wires, respectively, we introduce the ratio of the reaction rates per unit mass, i.e., the Cd ratio defined by the following equation (Cd ratio = $(RR)_{Cd}$), with the masses of the respective Au wires M_{bare} and M_{Cd}:

$$\text{Cd ratio} = (RR)_{Cd} = \frac{R_{bare}/M_{bare}}{R_{Cd}/M_{Cd}}. \tag{5.18}$$

5.5 Irradiation Experiments

Bare and Cd-tubed (see item (1) in Sect. 5.5.1) Au wires are attached to the core and irradiated to measure the relative spatial distributions of the activation reaction rates. The KUCA documents distributed on the day of the experiment shows the irradiation positions of the Au wires in addition to the core configuration. It is necessary to fully confirm the core arrangement and the irradiation positions of Au wires prior to the experiment.

5.5.1 *Gold Wires for Irradiation*

(1) Gold wire

To measure the relative spatial distributions of activation reaction rates, bare and Cd-tubed Au wires are set to the core. Prior to irradiation, a bare Au wire with a diameter of 0.5 mm and another Au wire inserted into a Cd tube (with a cavity of approximately 1.0 mm in diameter) with a thickness of 0.02 in. (approximately 0.51 mm), an outside diameter of 0.09 in. (approximately 2.29 mm) and an inside diameter of 0.05 in. (approximately 1.27 mm) are prepared. Both ends of the Cd tube should be securely sealed with radio pliers to prevent water from entering the tube during the irradiation. The length of the Au wire depends on the location where it is to be set, so confirm the length in advance.

(2) Experimental settings

When measuring the spatial distribution of the reaction rate in the vertical direction (direction z in Fig. 1.9) of the core, Al support plates fitted with bare and Cd-tubed Au wires are inserted between adjacent fuel frames from the top of the core and secured. To measure the spatial distributions of the reaction rates in directions x and y in Fig. 1.9, Al support plates fitted with bare Au wires are inserted between adjacent fuel frames and secured.

(3) Neutron irradiation

The Au wires are irradiated with neutrons while keeping the indicated value of the linear power meter Lin-N (UIC#5) at a constant value so that the reactor power is about 0.5 W (γ-ray dose rate: approximately 250 mR h^{-1}) as indicated by the dose rate on the γ-ray area monitor (γ AREA-C) in the C core (irradiation time: about 30 min). Then, the irradiation start time, irradiation end time, indicated value of the linear power meter during irradiation, dose rate indicated value of the γ-ray area monitor, control rod positions, core water level, and core temperature should be recorded. In addition, the positions of the detectors of the linear power meter and the γ-ray area monitor should be confirmed in advance.

(4) Removal after irradiation

After irradiation, when the radiation level in the C core decreases and access to the room is possible (usually the day after irradiation), the irradiated Au wires should be removed from the core. Since the dose rate in the upper part of the core is very high, it is necessary to carry a radiation detector to monitor the radiation dose carefully and remove the Au wires within a short time as much as possible. Special care should be taken to minimize the radiation dose not only when removing the Au wires, but also when transporting the Au wires to the radioactivity measurement room.

5.5.2 Measurements of Relative Reaction Rate Distributions by Gold Wires

(1) Cutting gold wire

We should cut the irradiated bare and Cd-tubed Au wires into pins of approximately 10 mm (or 15 mm) each in length. Also, we should insert each cut pin-shaped Au wire into a sample capsule, and number the capsules with the sample number so that the positions and order of the Au wires are correct. Then, it should be very careful not to mix the sample Au wire up when cutting the Au wire, and not to confuse the bare Au wire with the Cd-tubed Au wire. Furthermore, when inserting the cut pin-shaped Au wire into the sample capsule, we should make sure that the Au wire is placed to the bottom of the sample capsule and that the Au wire does not clump.[5]

(2) Measurement of radioactivity

As shown in Fig. 5.3, each sample capsule is placed in a designated polyethylene container and inserted into a well-type NaI(Tl) scintillation detector (or a well-type HPGe detector) to count the γ-ray at 411.8 keV. The counting is done by analyzing the γ-ray spectrum using a multi-channel analyzer (MCA). An example of range of interest (ROI) setting (MCA screen display) is shown in Fig. 5.4.

During the measurement, we should record the start time, measurement time, and count rate for each sample. Also, we should measure the background of the radiation detector and obtain the background count rate C_{BG}.

Using the background count rate C_{BG}, sequentially determine the average count rate $C'_{av,j}$ and net count rate $C_{av,j}$ ($= C'_{av,j} - C_{\mathrm{BG}}$) of the Au wire for each sample (subscript j denotes sample number).

(3) Weight of Au cutting sample

Using an electronic balance, measure and record the weight M_j of each sample pin (Au wire sample).

[5] The Au wire sample is placed in close contact with the bottom of the sample capsule to avoid the effects of changing geometric conditions, and the Au wire is not clumped to eliminate the effects of self-shielding of the γ-ray of 411.8 keV by the Au wire itself.

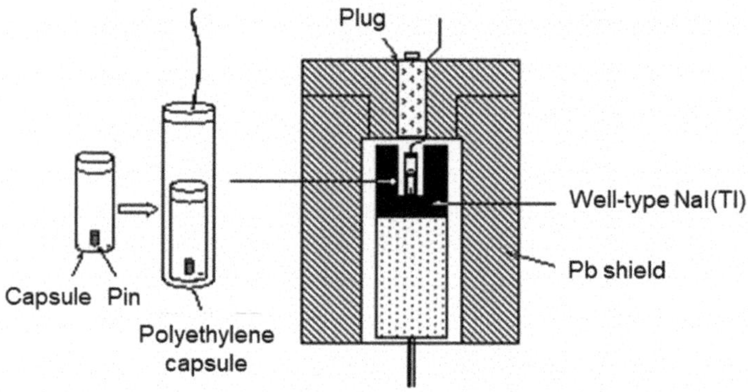

Fig. 5.3 Measurement image of radioactivity of Au wire pin using well-type NaI(Tl) scintillation detector (Ref. [4])

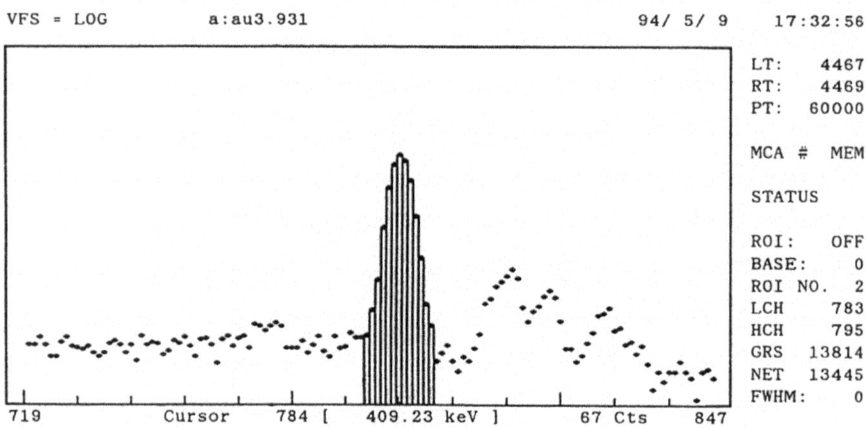

Fig. 5.4 Example of ROI setting and measurement of γ-ray peak counts (Ref. [4])

(4) Relative saturated activity

The value of the saturated activity can be obtained by substituting the net count rate $C_{av,j}$ of each sample into C_{av} on the right-hand side of Eq. (5.16). Note that since we are measuring the relative distribution of the reaction rate, Eq. (5.17) can be used here.

(5) Relative spatial distribution of reaction rates

Plotting the relative values $\left(R_{reaction,j}\right)_{\text{bare}}$ and $\left(R_{reaction,j}\right)_{\text{Cd}}$ of the respective reaction rates per unit mass for the sample pin positions of the bare and Cd-tubed Au wires gives the relative spatial distribution of the activation reaction rates corresponding to total neutron flux and epi-thermal neutron flux, respectively. By considering the correspondence between the irradiation positions of the bare and the Cd-tubed Au wires and subtracting the reaction rate per unit mass of the Cd-tubed Au wire from that of the bare Au wire, $(R_{reaction,j})_{\mathrm{bare}} - (R_{reaction,j})_{\mathrm{Cd}}$, the spatial distribution of reaction rate corresponding to thermal neutron flux can be obtained.

(6) Spatial distribution of Cd ratio in direction z

The spatial distribution of Cd ratio in the direction z (Fig. 1.9 in Chap. 1) can be obtained from $(R_{reaction,j})_{\mathrm{bare}}/(R_{reaction,j})_{\mathrm{Cd}}$ by considering the correspondence of irradiation positions.

5.6 Discussions

Experimental report is summarized in items (1)–(3) in Sects. 5.6.1 and 5.6.2 shows a list of experimental results of relative distributions of activation reaction rates by Au wires, in addition to discussion issues.

5.6.1 Experimental Reports

(1) Purpose of experiment
(2) Experimental procedures
(3) Experimental settings

- Name of core, core configuration and irradiation positions of Au wires (refer to relevant documents),
- Operation number (Run No.) during irradiation, positions of control rods and safety rods, water level, core temperature, indicated values of linear power meter and γ-ray area monitor, start time, and end time of Au wire irradiation,
- Type of γ-ray detector used to measure radioactivity.

5.6.2 Reaction Rate Distributions by Gold Wires

(1) Measured data of radioactivity

For the results of the radioactivity measurements of Au wires, consider the following items for each of bare and Cd-tubed Au wires:

- Number of Au wires,
- Locations of Au wires in the core,
- Start time of radioactivity measurement,
- Irradiation time,
- Waiting time,

- Measurement time,
- Counts and count rates for the γ-ray at 411.8 keV of all Au wire samples obtained by cutting,
- Background counts and count rates without any Au wire sample,
- Net count rates of the γ-ray at 411.8 keV (γ-ray count rates—background count rates) of all Au wire samples,
- Masses of all Au wire samples,
- Relative value of activation reaction rate shown in Eq. (5.17) of all Au wire samples.

(2) Discussions

- Compare the reaction rate distributions by the bare Au wires in directions x, y, and z (Fig. 1.9 in Chap. 1), and discuss the differences in shape and the peaks in the reflector regions relative to the core center.
- Describe reasons of the difference between the shapes of reaction rate distributions by bare and Cd-tubed Au wires in the direction z.
- Using the reaction rate distributions by bare and Cd-tubed Au wires in the direction z, show the relative spatial distributions of reaction rates corresponding to total and epi-thermal neutron fluxes in the fuel region in the direction z. Also, approximate the activation reaction rate corresponding to thermal neutron flux from the difference between the two reaction rate distributions.
- Using the reaction rate distributions by bare and Cd-tubed Au wires in the direction z, determine the Cd ratio of reaction rate distributions at the fuel region in the direction z with the use of Eq. (5.18). Also, from the viewpoint of neutron spectrum, discuss the differences in positions using the Cd ratio results in fuel and light-water reflector regions.

References

1. Gilmore G, Hemingway JD (1995) Practical gamma-ray spectrometry. John Wiley & Sons Ltd., Chichester, England
2. Firestone RB, Baglin CM, Chu SYF (eds) (1998) Table of isotopes, 8th edn. Wiley, New York
3. Shibata K, Iwamoto O, Nakagawa T et al (2011) JENDL-4.0: a new library for nuclear science and engineering. J Nucl Sci Technol 48:1–30. https://doi.org/10.1080/18811248.2011.9711675
4. Misawa T, Unesaki H, Pyeon CH (2010) Nuclear reactor physics experiments. Kyoto University Press, Kyoto, Japan. http://hdl.handle.net/2433/276400. Accessed 21 May 2024

Chapter 6
Determination of Neutron Flux

Abstract The physical quantity of neutron flux can be measured indirectly through interaction signals between the neutron and the material used as a neutron detector, as discussed in Chap. 5. To determine neutron flux, the measured reaction rate obtained in the experiments should be used. Acknowledging that it is not easy to determine neutron flux using the reaction rate, this chapter focuses on the theory of reactor physics related to neutron flux and discussion of radiation measurement techniques to determine neutron flux.

Keywords Thermal neutron flux · Radioactivity · Neutron spectrum · Perturbation effect · Reactor power

6.1 Objectives

In learning how to determine neutron flux presented in this chapter, keep in mind the objectives of the experiments described below. Noteworthy is that the relationship between reaction rate and neutron flux could help you better understand the importance of the method of determining the neutron flux. The main objectives of the experiments are as follows:

- To determine the absolute thermal neutron flux in the core and the reactor power using the activation of gold-197 (^{197}Au) foils.
- To gain insight into the neutron spectrum in the core using the results of reaction rate obtained from bare and cadmium (Cd)-covered Au foils.

C. H. Pyeon et al., *Reactor Laboratory Experiments at Kyoto University Critical Assembly*, https://doi.org/10.1007/978-981-97-8070-9_6

Table 6.1 Change in neutron temperature T_n of non-$1/\upsilon$ correction factors $g(T_n)$ for ^{197}Au

T_n (K)	$g(T_n)$
293	1.0053
313	1.0064
333	1.0075
353	1.0086
373	1.0097

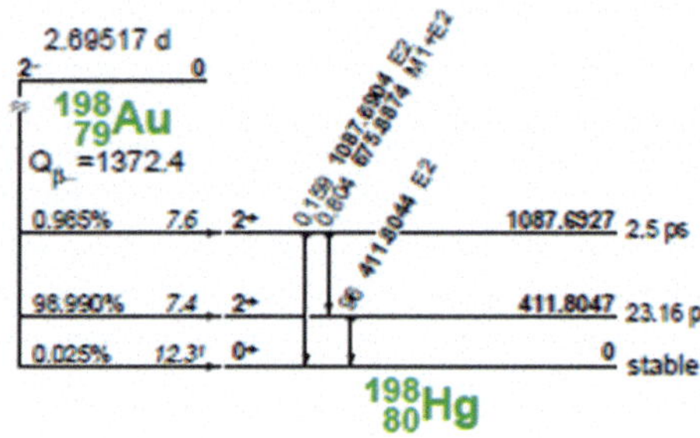

Fig. 6.1 Decay scheme of ^{198}Au (data from Table of isotopes in Ref. [2])

6.2 Measurements of Radioactivity

6.2.1 Principle

In reactor physics experiments, ^{197}Au is the most widely used as an activation foil for obtaining information on thermal neutron flux: the thermal neutron flux distribution in the core, the absolute value of thermal neutron flux, and the reactor power. In the experiments presented in this section, the Au foil is also used to measure the absolute value of thermal neutron flux. The following method is used to measure the average count rate of induced radioactivity C_{av} shown in Eq. (5.14) for the Au activation foil. The reasons why the Au is often used as an activation foil for measuring thermal neutron flux in reactor physics experiments are as follows:

- It is known that the Au consists of ^{197}Au in the abundance ratio 100%, and the energy dependence of the neutron-induced reaction crosse section ^{197}Au (n, γ)^{198}Au (capture reactions)[1] follows approximately the $1/\upsilon$ law in the thermal neutron region.
- It is well known that the activation cross sections of ^{197}Au are $\sigma_{\mathrm{act}}(\upsilon_0) = 98.5 \pm 0.4$ barn (Ref. [1]) for neutrons with neutron velocity $\upsilon_0 = 2200$ m s^{-1}.
- The decay scheme of ^{198}Au induced by ^{197}Au activation is relatively simple, as shown in Fig. 6.1, and its decay rate is easy to determine.
- The half-life of ^{198}Au is 2.69517 days, which is suitable for measuring radioactivity.

[1] See Fig. 6.4. The non-$1/\upsilon$ correction factor $g(T_{\mathrm{n}})$ in Table 6.1 is also approximately unity.

As shown in Fig. 6.1, the induced nuclei ^{198}Au produced by ^{197}Au activation turn into mercury-198 (^{198}Hg) by β^--decay. The amount of about 99% of the nuclei decays with the emittance of β-ray of maximum energy $E_{\beta\max} = 961$ keV, and goes then to the ground state after emitting γ-ray of 411.80 keV through the first excited state of ^{198}Hg. Since the half-life of this first excited state is as short as about 23 ps, it can be assumed that the β-ray and γ-ray are emitted simultaneously with the transition from ^{198}Au to ^{198}Hg. The $4\pi\beta - \gamma$ coincidence method is an absolute measurement method for activation of the Au foil using the simultaneous emission of the β-ray and γ-ray. See Sect. 6.8 for more information on the coincidence method.

Since the high-purity germanium (HPGe) detector with excellent energy resolution has been widely used, the decay rate of γ-ray emitting nuclei or the emission rate of γ-ray photons is obtained using HPGe detectors with pre-calibrated detection efficiency. The method of considering it as the absolute value of the amount of activation has been widely used.[2] Although this method is not strictly an "absolute measurement," its use is spreading because of its simpler configuration and adjustment of the measurement circuit system including detectors compared to the $4\pi\beta - \gamma$ coincidence method. In this experiment, the HPGe detector will be used for absolute measurement of the activity of Au foil. In the relative measurement of the total activation reaction rate using the Au wire, the sodium iodide (thallium) (NaI(Tl)) detector, which is simpler to adjust the detector, can also be used.

In measurements using an HPGe detector or NaI(Tl) detector, a multi-channel analyzer (MCA) is used to measure the pulsed-height distribution, and the full energy absorption peak counts for the γ-ray at 411.80 keV are obtained. The basic method is to quantify the radioactivity intensity by determining the number of total energy absorption peaks for γ-ray at 411.80 keV. The details of the measurement will be described in Sect. 6.5.

The following relationship exists between the count rate C_p (s^{-1}) of the measured full energy absorption peaks and the decay rate D (s^{-1}) of the irradiated Au foil[3]:

$$C_p = \frac{\varepsilon_{\mathrm{D}} \varepsilon_{\mathrm{E}} D}{1 + \alpha}, \tag{6.1}$$

where, α indicates the internal conversion factor ($\alpha = 0.041$ in Ref. [3]). Also, ε_{D} and ε_{E} are the detection efficiencies with respect to the full energy absorption peaks and the emission rate of γ-ray energy, respectively. To obtain the absolute value of

[2] Although similar measurements can be made with NaI(Tl) scintillation detectors, the energy resolution is significantly poorer than that of semiconductor detectors, resulting in large errors in determining the full energy absorption peak counts, making it difficult to obtain high measurement accuracy. Therefore, it is difficult to obtain high accuracy of measurements. NaI(Tl) scintillation detectors are used exclusively for relative measurements, while semiconductor detectors are used for absolute measurements. Note that this method is not strictly an absolute measurement, since a standard source should be used to determine the detection efficiency.

[3] Simplifying the decay scheme shown in Fig. 6.1, and ignoring β-decay to the ^{198}Hg ground level directly by β-decay and decay to the 1088 keV level (to be precise, 18% of those decaying to the 1088 keV level decay to the ground level without going through the first excited level of ^{198}Hg), all of them emit the γ-ray of 411.80 keV via the first excited level of ^{198}Hg.

D, the detection efficiency ε_D with respect to the full energy absorption peaks of the γ-ray to be detected should be determined.

6.2.2 Detection Efficiency

The detection efficiency ε_D should be obtained using a standard source with the calibrated intensity of the source. If the standard source is calibrated by the number of γ-ray photons emitted, it is obtained by Eq. (6.2); if the standard source is calibrated by the decay rate, it is obtained by Eq. (6.3), as follows:

$$\varepsilon_{pi} = \frac{C_{pi}}{Q_i}, \tag{6.2}$$

$$\varepsilon_{pi} = \frac{C_{pi}}{F_i D_i}, \tag{6.3}$$

where C_{pi} is the count rate of total energy absorption peaks for γ-ray at energy E_i, Q_i and D_i are the γ-ray emission rate and decay rate [Bq] of the standard source at the time of measurement, respectively. Also, F_i is the ratio of γ-ray at energy E_i,[4] and the subscript i represents the γ-ray energy that was measured.

The detection efficiency depends on the energy of the γ-ray to be detected. If the detection efficiency for the γ-ray at 411.80 keV cannot be determined directly, it is necessary to determine the detection efficiency ε_{pi} for each of the γ-ray using standard sources emitting γ-ray with energies around that value, and to determine ε_{pi} as a function of E_i by the least-squares method or ε_D by the interpolation method for the γ-ray at 411.80 keV.

The types of standard radiation sources are available as follows:

(1) Mixed-nuclei source

The mixed-nuclei source is covering with the energy range between 65 keV and 1.8 MeV. Also, the source is very convenient to use because it is possible to calibrate the energy and detection efficiency over the entire energy range in a single measurement. However, since the half-life of each nuclide is different, it is necessary to correct for each activity.

(2) Single-nucleus source

The use of single-nucleus source is the most common method of correcting for the required energy range using several different sources. When this method is used to correct detection efficiency, special attention should be paid to the reproducibility of the position of each source. The detection efficiency depends on the geometrical conditions of the detector and the source in addition to the γ-ray energy. The geometrical conditions should be the same as possible, when determining the detection

[4] It can be obtained from decay diagrams of nuclides used as standard sources.

efficiency using a standard source and measuring irradiated samples.[5] In particular, when measuring the radioactivity of a large number of samples to obtain a spatial distribution, special care should be taken to ensure that the detection efficiency does not change during the measurement of each sample.

6.3 Effects of Neutron Spectrum

As is clear from Eq. (5.16), the total activation reaction rate is the reaction rate by neutrons of all energies, and information on the thermal neutron flux, which is the target of the measurement, is buried in the total activation reaction rate. A method for extracting this buried information is presented in this section. In this section, neutron spectrum in thermal neutron reactors is described, and thermal neutrons, epi-thermal neutrons and neutron spectra are introduced. Next, the activation reaction rate by thermal neutron flux is explained, and the experimental devices are introduced, to obtain the thermal neutron flux from the activation reaction rate and various corrections based on the theory of reactor physics.

In thermal neutron reactors, the neutron spectrum can be divided into three regions according to energy as follows:

(i) Fission energy,
(ii) Moderation energy,
(iii) Thermal energy.

Region (i) is formed as a fission spectrum when fission neutrons produced in the fuel assembly, and both prompt and delayed neutrons have kinetic energy ranging from about 75 keV–17 MeV with the average of approximately 1 MeV. When the fission neutrons are moderated in the core medium, especially by interaction with lighter nuclei in the moderator, region (ii) is formed; its neutron energy range is approximately $1\,\text{eV} < E < 100\,\text{keV}$ and, according to moderation theory, the energy change in the neutron flux is proportional to $1/E$. In the last process of moderation, the neutrons reach a state of thermal equilibrium with the medium atoms. In this state, region (iii) is formed, and the neutron energy range is approximately $E < 1\,\text{eV}$.

Regions (ii) and (iii) are particularly important for thermal neutron reactors. For simplicity, let us assume that the medium in the core has an infinite and uniformly distributed fast neutron source. If the medium is an ideal moderator with large scattering cross sections and zero absorption cross sections, in region (iii), the energy distribution of the thermal neutron flux (thermal neutron spectrum) is then a Maxwellian distribution determined by the physical temperature T_m (K) of the

[5] Although the Au foil and wire are considered an area source and a line source, respectively, most standard sources are closer to point sources. By separating a certain type of line source, the activation foil, from the detector, it is possible to consider the foil as a point source. However, such measurements generally result in lower count rates and lower statistical accuracy. For example, to improve the accuracy of absolute measurement, it is desirable to make the geometry of the standard source the same as that of the activation foil to be measured.

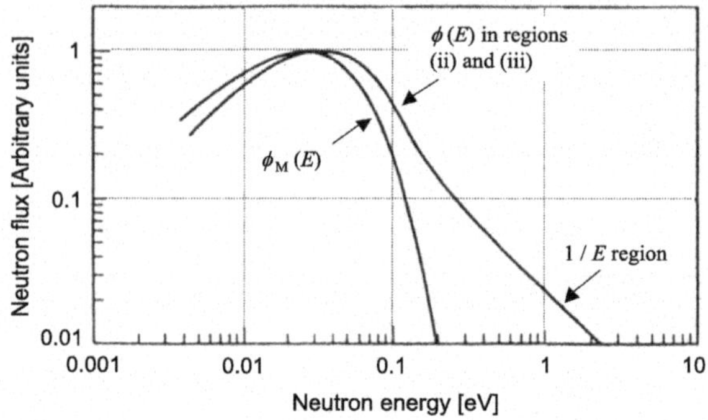

Fig. 6.2 Conceptual diagram of neutron spectrum $\phi(E)$ in a thermal reactor and the Maxwellian thermal neutron spectrum $\phi_M(E)$ in a pure moderator (both are normalized to a maximum value of unity) (Ref. [4])

medium. The thermal neutron flux $\phi_M(E)$ ($cm^{-2}\,s^{-1}\,MeV^{-1}$) per unit energy width at energy E can be expressed as follows:

$$\phi_M(E) \propto E(kT_m)^{-2}e^{-\frac{E}{(kT_m)}}, \tag{6.4}$$

where k is the Boltzmann constant.

A conceptual diagram of the energy distribution of the thermal neutron flux is shown in Fig. 6.2. Here, the energy with maximum $\phi_M(E)$ reaches a maximum E_p(eV) is as follows:

$$E_p = kT_m. \tag{6.5}$$

For example, when T_m is 293.6 K, $E_p = 0.0253$ eV with the Boltzmann constant $k = 8.6167 \times 10^{-5}$ eV K^{-1} and $E_p = 0.0253$ eV (the velocity corresponding to E_p is called the most probable velocity, $\upsilon_p = 2200$ m s^{-1}). However, in an actual nuclear reactor, it is impossible to regard the medium as a pure moderator because it contains materials with large neutron absorption, such as nuclear fuel. Therefore, when regions (ii) and (iii) are displayed together, their neutron spectra can be classified into the following three parts, as shown in Fig. 6.2.

- $E \leq 0.1$ eV: thermal energy region that can be approximated by the Maxwellian distribution corresponding to the temperature T_n (termed the neutron temperature) that is higher than T_m due to absorption effects (region (iii)),
- 0.1 eV $< E <$ 1 eV: transient regions between regions (ii) and (iii),
- $E >$ 1 eV:$1/E$ law region (region (ii)).

The region (ii) and the transient region are sometimes combined into one region, termed the "epi-thermal" region, which is distinguished from the thermal region in region (iii).

From the above description, the spectrum of a thermal neutron reactor can be written as $\phi(E)$, using the energy-dependent neutron flux per unit energy width as follows (Refs. [5, 6]):

$$\phi(E) = \phi_{\text{th}} E (kT_n)^{-2} e^{-\frac{E}{(kT_n)}} + \phi_{\text{epi}} \frac{\Delta\left(\frac{E}{kT_n}\right)}{E}, \tag{6.6}$$

where the coefficient ϕ_{th} in the first term on the right-hand side in Eq. (6.6) is the thermal neutron flux. The coefficient ϕ_{epi} in the second term of the right-hand side is the neutron flux in the epi-thermal region (more precisely, the neutron flux per lethargy). The neutron temperature T_n in Eq. (6.6) is difficult to define clearly and various tabular expressions have been proposed. For simplicity, an equation based on the heavy gas model (Ref. [4]) is given as follows:

$$T_n = \frac{T_m}{1 - \frac{\sum_a(T_n)}{\xi \sum_s}}, \tag{6.7}$$

where $\sum_a$ indicates the macroscopic absorption cross section of the medium, ξ the logarithmic average energy loss per collision, $\sum_s$ the macroscopic scattering cross section of the medium, and $\xi \sum_s$ the average slowing-down power of the medium in the core for epi-thermal neutrons. The coupling function $\Delta\left(\frac{E}{kT_n}\right)$ on the right-hand side of Eq. (6.6) is the function introduced to represent the neutron spectrum in the transient region. As shown in Fig. 6.3, the coupling function $\Delta\left(\frac{E}{kT_n}\right)$ shows somewhat different energy variations depending on the value of the moderating ratio $\frac{\xi \sum_s}{\sum_a(T_n)}$ of the medium, but is zero less than about $4kT_n$, maximum around $8kT_n$, and unity more than $15kT_n$ (Ref. [6], Fig. 6.4).

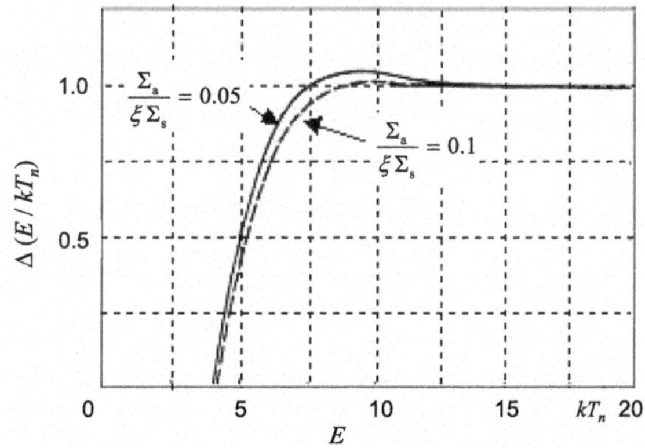

Fig. 6.3 Energy dependence of coupling function $\Delta\left(\frac{E}{kT_n}\right)$ (Ref. [4])

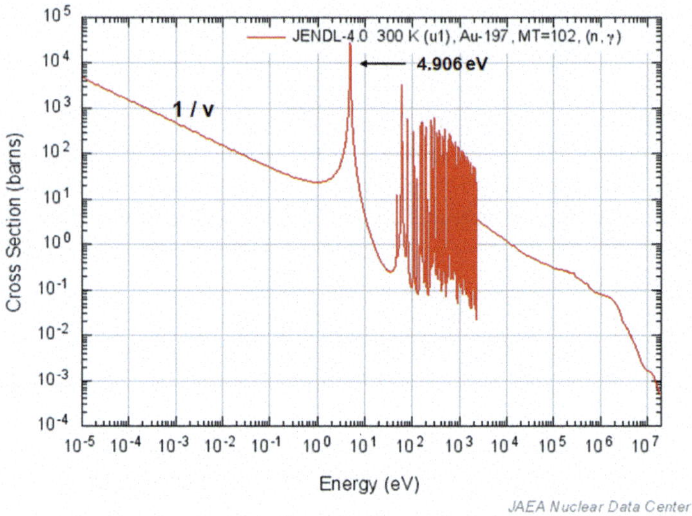

Fig. 6.4 ^{197}Au capture cross sections (data from JENDL-4.0 in Ref. [8])

6.4 Determination of Thermal Neutron Flux

If the upper energy limit of thermal neutrons is E_{th}, the number of atoms in the activation foil is N, the activation cross section is $\sigma(E)$ (cm^2), and the energy dependent neutron flux is $\phi(E)dE$ (cm^{-2} s^{-1}), the activation reaction rate R_{th} (s^{-1}) with thermal neutrons can be expressed as in Eq. (5.3), as follows:

$$R_{th} = \int_0^{E_{th}} N\sigma(E)\phi(E)dE = N\sigma_{\text{act}}\phi_{\text{th}}, \tag{6.8}$$

where ϕ_{th} is the total thermal neutron flux and σ_{act} the average activation cross section for thermal neutrons, which can be expressed as follows, respectively:

$$\phi_{\text{th}} = \int_0^{E_{th}} \phi(E)dE, \tag{6.9}$$

$$\sigma_{\text{act}} = \int_0^{E_{th}} \frac{\sigma(E)\phi(E)dE}{\phi_{\text{th}}}. \tag{6.10}$$

Although the activation reaction rate due to thermal neutrons R_{th} in Eq. (6.8) and the total activation reaction rate R_{act} in Eq. (5.3) appear to differ only in the upper limit of the energy integral in the mathematical formula, various devices and corrections described in Sects. 6.4.1 through 6.4.4 are necessary to take this difference into account and obtain information on the thermal neutron flux from the experiment.

6.4.1 Average Activation Cross Section for Thermal Neutrons

To find the total thermal neutron flux ϕ_{th} from the activation reaction rate R_{th} by thermal neutrons, we need to know the value of σ_{act}. For activation foil materials in which the energy variation of the activation cross section follows the $1/\upsilon$ law, the value of σ_{act} can be obtained by correcting the activation cross section $\sigma(\upsilon_{\text{p}})$ corresponding to the velocity $\upsilon_{\text{p}} = 2200\ \text{m s}^{-1}$ for the average energy of the Maxwellian distribution with neutron temperature T_n. However, for general activation materials that do not follow the $1/\upsilon$ law, this correction is not sufficient and should be further corrected using the non-$1/\upsilon$ correction factor $g(T_n)$ to obtain σ_{act} as follows:

$$\sigma_{\text{act}} = g(T_n)\sqrt{\frac{293.6\pi}{T_n}}\frac{\sigma(\upsilon_{\text{p}})}{2}. \tag{6.11}$$

Table 6.1 shows the results of calculating the non-$1/\upsilon$ correction factor $g(T_n)$ as a function of T_n for ^{197}Au (see Ref. [7]).

6.4.2 Correction for Contribution by Thermal Neutrons

The value of R'_{th} in Eq. (5.19) is the activation reaction rate obtained by neutrons with energies lower than E_{Cd}. For $E_{\text{Cd}} > E_{\text{et}}$ (Fig. 5.1), to obtain the contribution of thermal neutrons lower than E_{et}, the correction should be made to R'_{th} for the contribution due to neutrons with energies at $E_{\text{et}} < E < E_{\text{Cd}}$.[6]

For this purpose, we will use the correction factor F_{Cd} in Eq. (6.12). This is a quantity defined as the ratio of the activation reaction rate R_{epi} produced in the Au foil by epi-thermal neutrons ($E \geq E_{\text{et}}$), because of the neutron flux with the $1/E$ spectrum ($\phi(E) \propto 1/E$), comparing with the activation reaction rate $R_{\text{Cd},epi}$ produced by neutrons with energies above the effective Cd cut-off energy E_{Cd} ($E \geq E_{\text{Cd}}$). Here,

[6] The thermal spectrum cut-off energy, the boundary energy between the thermal and epi-thermal regions is $E_{\text{et}} \approx 0.1$ eV ($< E_{\text{Cd}}$), but E_{et} is also not completely defined. When determining E_{et}, we also approximate the energy change of the coupling function $\Delta(E/kT_n)$ in a stepwise manner in Fig. 6.3, and assume $\Delta(E/kT_n) = 0$ at $E \leq E_{\text{et}}$ and $\Delta(E/kT_n) = 1$ at $E > E_{\text{et}}$. Then, as in the case of E_{Cd}, the activation reaction rate from the neutron spectrum of the complete $1/E$-type distribution at $E_{\text{et}} < E < \infty$ is defined as E_{et}, as is equal to the activation reaction rate from the neutron spectrum of the $1/E$-type distribution with the coupling function at $0 \leq E < \infty$. For very thin $1/\upsilon$-type activation foils, it is $E_{\text{et}} = \mu k T_n$. For example, it is $\mu = 3.6 \pm 0.4$ when the medium is heavy water, and $\mu = 3.4 \pm 0.3$ for graphite. Also, for $kT_n = 0.0253$ eV, it is $E_{\text{et}} = 0.09$ eV for heavy water and $E_{\text{et}} = 0.086$ eV for graphite, and both are very close to 0.1 eV (Ref. [6]).

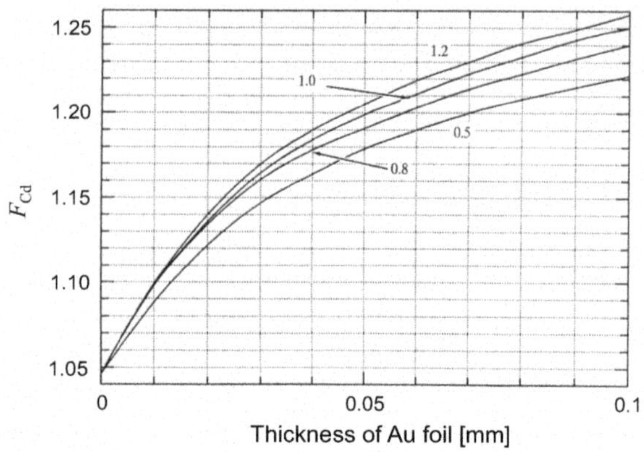

Fig. 6.5 Dependence of F_{Cd} on thickness of Au foil (Ref. [4]) (the number in each curve represents the thickness (mm) of the Cd cover.)

the value of F_{Cd} is expressed as follows:

$$F_{Cd} = \frac{R_{epi}}{R_{Cd,epi}} = \frac{\int_{E_{et}}^{\infty} \sigma_{act}(E)dE/E}{\int_{E_{Cd}}^{\infty} \sigma_{act}(E)dE/E} = 1 + \frac{\int_{E_{et}}^{E_{Cd}} \sigma_{act}(E)dE/E}{\int_{E_{Cd}}^{\infty} \sigma_{act}(E)dE/E}. \tag{6.12}$$

From Eq. (6.12), it is clear that F_{Cd} is larger than unity. Note that F_{Cd} for a very thin $1/\upsilon$-type activation foil can be expressed as follows (see Ref. [6]):

$$F_{Cd} = \sqrt{\frac{E_{Cd}}{E_{epi}}}, \tag{6.13}$$

where E_{epi} represents the energy of epi-thermal neutrons. For example, a thin Cd cover with 1 mm thick has $F_{Cd} \approx 2.75$ at $T_n = 293.6$ K. The ^{197}Au activation cross section has, however, a strong resonance peak at 4.906 eV with a higher energy position than E_{Cd} (0.5 eV), as shown in Fig. 6.4. Consequently, the value of F_{Cd} takes a lower order value depending on both Au foil and Cd cover thickness, as shown in Fig. 6.5.

6.4.3 Correction for Thermal Neutron Transmittance in Cadmium Cover

The Cd cover is used to prevent thermal neutrons from being absorbed by the activation foil. However, as shown in Fig. 6.6, the effect of thermal neutron transmission in the Cd cover cannot be ignored if there is a small ratio of thermal neutron transmission in the Cd cover, i.e., the thermal neutron transmittance $(TR)_{Cd}$, when the Cd cover is thin. Considering three factors, the Cd ratio ($(RR)_{Cd}$), F_{Cd}, and the thermal neutron transmittance $(TR)_{Cd}$ with respect to the activation reaction rate R_{bare} of the

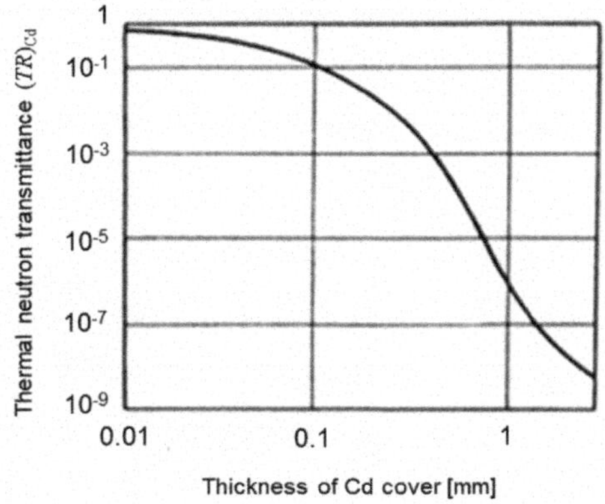

Fig. 6.6 Dependence of thermal neutron transmittance on Cd cover thickness (Ref. [4])

bare Au foil, the activation reaction rate R''_{th} with respect to the thermal neutron flux can be expressed as follows:

$$R''_{th} = \frac{R_{bare}\left(1 - \frac{F_{\mathrm{Cd}}}{(RR)_{\mathrm{Cd}}}\right)}{1 - F_{\mathrm{Cd}} \cdot (TR)_{\mathrm{Cd}}}. \tag{6.14}$$

6.4.4 *Perturbation Effect of Thermal Neutron Flux*

By combining $(RR)_{\mathrm{Cd}}$, F_{Cd}, and $(TR)_{\mathrm{Cd}}$ for the activation reaction rate of bare Au foil R_{bare}, the activation reaction rate by thermal neutrons in the Au foil can be determined. In the following, we will discuss the relationship between the activation reaction rate by thermal neutrons and the thermal neutron flux of the system.

In general, when an activation foil is set at the place with thermal neutron flux ϕ_{th}, the thermal neutron flux distribution inside and around the foil is perturbed and decreases as shown in Fig. 6.7. The phenomenon is called the perturbation, which consists of two effects: the reduction of thermal neutron flux around the foil (depression) and the further reduction of neutron flux inside the foil due to distortion (self-shielding). The two effects are collectively termed the perturbation effect.

To express the perturbation effect, we will use the ratio of the average thermal neutron flux inside the activation foil $\overline{\phi_{\mathrm{th}}}$ to the thermal neutron flux before disturbance ϕ_{th}. The ratio is called the perturbation factor and is denoted by f as follows:

$$f = \frac{\overline{\phi_{\mathrm{th}}}}{\phi_{\mathrm{th}}}. \tag{6.15}$$

Fig. 6.7 Perturbation effect of thermal neutron flux (Ref. [4])

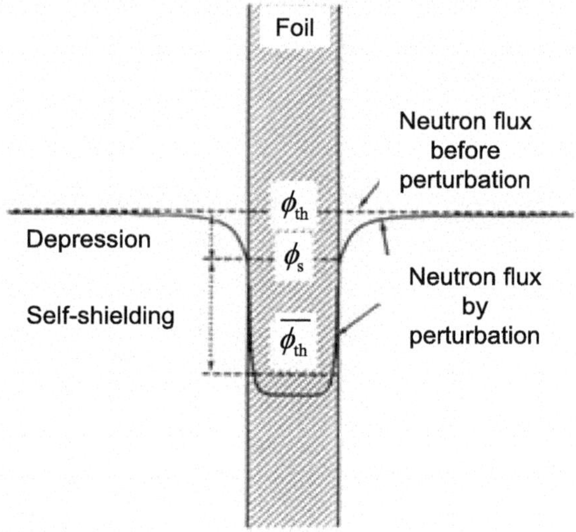

Furthermore, let ϕ_s the thermal neutron flux at the foil surface when the foil is inserted, and the perturbation factor f in Eq. (6.15) be divided into a depression factor $f_1 = \frac{\phi_s}{\phi_{th}}$ and a self-shielding factor $f_2 = \frac{\overline{\phi_{th}}}{\phi_s}$ as follows:

$$f = \frac{\phi_s}{\phi_{th}} \frac{\overline{\phi_{th}}}{\phi_s} = f_1 f_2. \tag{6.16}$$

Here, as shown in Fig. 6.7, all of these factors are less than unity.

The R''_{th} in Eq. (6.14) is caused by the average thermal neutrons $\overline{\phi_{th}}$ inside the activation foil, and the R_{th} in Eq. (6.8) is caused by the undisturbed original thermal neutron flux ϕ_{th}. The relationship between R_{th} and R''_{th} can be then expressed as follows:

$$R_{th} = R''_{th}\left(\frac{\phi_{th}}{\overline{\phi_{th}}}\right) = \frac{R''_{th}}{f}. \tag{6.17}$$

From Eq. (6.17), the desired value of R_{th} can be obtained by correcting the value of R''_{th} from the measurement with the perturbation factor f.

When measuring the neutron flux using the ^{197}Au activation foil in a core consisting of the KUCA-C core, the depression factor f_1 in the core can be regarded as $f_1 \approx 1$. So, Eq. (6.16) can be regarded as $f \approx f_2$.

The self-shielding factor f_2 can be expressed as follows: if the macroscopic absorption cross section of thermal neutrons in an activation foil with thickness t (cm) is $\sum_a$(cm^{-1}) and the thickness expressed in units of the absorption mean free path λ_a (cm) is $\sum_a t$ $(=\frac{t}{\lambda_a})$, the perturbation factor f can be expressed as follows:

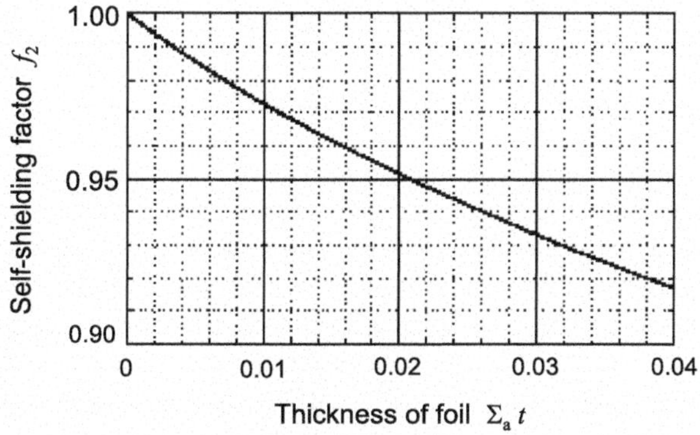

Fig. 6.8 Self-shielding factor for isotropic incidence of thermal neutrons on an infinite flat foil f_2 (Ref. [4])

$$f \cong f_2 = 1 - \frac{2E_3(\sum_a t)}{(2\sum_a t)}, \tag{6.18}$$

where E_3 is the third-order exponential integral, and detailed values are given in Ref. [6]. An example calculation of f_2 is shown in Fig. 6.8. The details of f_1 and f_2 are given in Ref. [9].

From the above results, substituting Eqs. (6.14) and (6.17) into Eq. (6.8), the thermal neutron flux can be finally expressed as follows:

$$\phi_{\text{th}} = \frac{R_{th}}{N\sigma_{\text{act}}} = \frac{R_{bare}\left(1 - \frac{F_{\text{Cd}}}{(RR)_{\text{Cd}}}\right)}{N\sigma_{\text{act}} f\,(1 - F_{\text{Cd}}(TR)_{\text{Cd}})}. \tag{6.19}$$

To obtain the absolute value of the thermal neutron flux using Eq. (6.19), the absolute value of R_{bare} should be determined. Furthermore, the absolute value of the decay rate of the activation foil should be measured. The measurement of the absolute value is performed using HPGe detectors. For details on the measurement technique, refer to Sect. 6.5.

6.5 Irradiation Experiments

In the determination of neutron flux, bare and Cd-covered Au foils are attached to suitable positions of the core and irradiated at representative positions in the core, to obtain absolute values of thermal neutron flux. The KUCA documents distributed on the day of the experiment show the irradiation positions of the Au foils in addition to the core configuration. It is necessary to fully confirm the core arrangement and the irradiation positions of Au foils prior to the experiment.

6.5.1 Gold Foils for Irradiation

(1) Gold foil

To determine the absolute thermal neutron flux at representative locations in the core, bare and Cd-covered Au foils are irradiated in the core. Prior to irradiation, a bare Au foil with a diameter of 8 mm and a thickness of 0.05 mm is wrapped in Al foil and another Au foil is enclosed using two special Cd covers with a diameter of 10 mm and a thickness of 0.8 mm. Each is then attached directly to the fuel plate. After the Cd-covered Au foil is placed in the cover body, a thin vacancy between two Cd covers should be securely sealed with tape or other means to prevent water from entering the Cd cover during irradiation.

(2) Experimental settings

Bare and Cd-covered Au foils are attached directly to the fuel plate and the Au foil-attached fuel plate is returned to its proper position on the fuel frame.

(3) Neutron irradiation

The Au foils are irradiated with neutrons while keeping the indicated value of the linear power meter Lin-N (UIC#5) at a constant value so that the core power is about 0.5 W (γ-ray area monitor: approximately 250 $\mathrm{mR\,h^{-1}}$) as indicated by the dose rate on the γ-ray area monitor (γ AREA-C) in the C core (irradiation time: about 30 min). At this time, the irradiation start time, irradiation end time, indicated value of the linear power meter during irradiation, dose rate indicated value of the γ-ray area monitor, control rod positions, core water level, and core temperature should be recorded. In addition, the positions of the detectors of the linear power meter and the γ-ray area monitor should be confirmed in advance.

(4) Removal after irradiation

After irradiation, when the radiation level in the C core decreases and access to the room is possible (usually the day after irradiation), the irradiated Au foils should be removed from the core. Since the dose rate in the upper part of the core is very high, it is necessary to carry a radiation detector to monitor the radiation dose carefully and remove the Au foils within a short time as possible. Special care should be taken to minimize the radiation dose not only when removing the Au foils, but also when transporting the Au foils to the radioactivity measurement room.

6.5.2 Measurements of Reaction Rate Using Gold Foil

See items (2) and (4) in Sect. 5.5.2, Chap. 5.

6.5.3 *Measurements of Radioactivity Using Gold Foil*

In measuring the radioactivity of Au foil and determining the thermal neutron flux, it is necessary to determine the absolute value of the radioactivity. Here, the procedure for measurement using the HPGe detector is described. The basic procedure is the same as that for measuring the radiation by the Au wire, but note that it is necessary to determine the detection efficiency to obtain the absolute value of radioactivity.

(1) Determination of detection efficiency for γ-ray at 411.8 keV

- Set up the measurement system according to handling precautions.
- Check the MCA's γ-ray spectrum display screen and adjust the gain of the main amplifier so that the required energy range can be measured using the specified standard γ-ray source. Plot the relationship between the central channel of the peak and γ-ray energy on a graph to eliminate misidentification of the peak.
- For all indicated γ-ray, find the counts of total energy absorption peaks (sometimes called peak area); while checking MCA's γ-ray spectrum display screen, move the cursor near the desired peak, zoom in on the spectrum as needed, make a range of interest (ROI) at the peak, and set the ROI in the area of the peak. By such an operation, many MCAs display the total counts for the area in the ROI and the net counts for the peak area from which the continuous component is excluded. If you use an MCA that does not have such a function, you should plot the spectrum near the peak on graph paper and determine the peak counts.
- Using Eqs. (6.2) or (6.3), calculate the detection efficiency ε_i for each energy E_i, plot the relationship between energy and full energy absorption detection efficiency on a double logarithmic graph, and perform first-order fitting (or interpolation) to determine the detection efficiency ε_D for the γ-ray at 411.8 keV.

(2) Measurement of radioactivity by gold foil

Place the irradiated Au foil in the same position as the standard source and measure the γ-ray at 411.8 keV using the same procedure with the standard source measurement. Then, the measurement start time and measurement time should be recorded. Then, from the count rate of the full energy absorption peak, calculate the saturated activity D_∞ using Eq. (5.15) (see Chap. 5).

(3) Weight of gold foil

Using an electronic balance, measure and record the weight M of the Au foil. This allows the number of Au atoms N in the Au foil to be determined.

6.5.4 Determination of Thermal Neutron Flux

We describe how the absolute value of the thermal neutron flux is determined using bare and Cd-covered Au foils. To apply the weight correction of the sample and the time decay correction of radioactivity to the measured saturated activity D_∞, the absolute value of the activated reaction rate is obtained by substituting M in Sect. 6.5.3 into Eq. (5.16). Then, based on the absolute value of the activated reaction rate for the bare Au foil, the thermal neutron flux ϕ_{th} at the measurement position is determined using Eq. (6.19).

The Cd ratio $(RR)_{\text{Cd}}$, F_{Cd}, thermal neutron transmittance $(TR)_{\text{Cd}}$, perturbation factor f, and average activation cross section for thermal neutrons of the activation foil σ_{act} in Eq. (6.19) are determined as follows:

(1) Cadmium ratio

The Cd ratio is obtained, which is the ratio of saturated activities per unit mass, from the following equation:

$$(RR)_{\text{Cd}} = \frac{(R_{\text{reaction}})_{\text{bare}}/M_{\text{bare}}}{(R_{\text{reaction}})_{\text{Cd}}/M_{\text{Cd}}}. \tag{6.20}$$

(2) F_{Cd}

The value of F_{Cd} is obtained from the thickness of the Au foil used and the thickness of the Cd cover using Fig. 6.5.

(3) Thermal neutron transmittance $(TR)_{\text{Cd}}$

The value of $(TR)_{\text{Cd}}$ is obtained from the thickness of the Cd cover using Fig. 6.6.

(4) Average activation cross section of activation foil for thermal neutrons σ_{act}

- As shown in Eq. (6.11), σ_{act} is a function of the neutron temperature T_n, so T_n is obtained using Eq. (6.7). The required medium temperature T_m, absorption cross section $\sum_{\text{a}}(T_n)$, and slowing-down power $\xi\sum_{\text{s}}$ are obtained as follows:
- The medium temperature T_m shall be equal to the temperature of the core.
- The absorption cross section $\sum_{\text{a}}(T_n)$ is a function of the unknown T_n, but here, as a first approximation, it is equal to $\sum_{2\text{a}}$ used in two-energy-group diffusion calculations, i.e.$\sum_{\text{a}}(T_n) \approx \sum_{2\text{a}}$. Note that the values of $\sum_{2\text{a}}$ are different depending on the fuel frame (Table 1.2 in Chap. 1).
- The slowing-down power $\xi\sum_{\text{s}}$ is considered to be the unit cell of the homogenized fuel assembly, and the core average slowing-down power $(\xi\sum_{\text{s}})_{\text{core}}$ is calculated from the volume ratio of each component material (uranium: U, light-water: H_2O and aluminum: Al) and the slowing-down power of each component material, which is used as the slowing-down power $\xi\sum_{\text{s}}$. The values shown in Table 6.2 are used for the slowing-down power of each component material.

Table 6.2 Slowing-down power $\xi \sum_s$ of core component materials (U, H_2O, and Al)

Component material	Slowing-down power $\xi \sum_s$ (cm^{-1})
U	0.003
H_2O	1.35
Al	0.0061

- The volume ratio of each component material (U, H_2O, and Al) is determined by considering a horizontal cross section of the unit cell of fuel assembly, 142 mm long and 71 mm wide, containing the fuel frame crossing the fuel meat section, indicating Figs. 1.6 (fuel frame) and 1.7 (unit cell of fuel assembly), and Table 1.1 in Sect. 1.2.2 in Chap. 1. Approximate the area ratio of each component material to be equal to that of each component material in that cross section.
- The volume ratio of U to Al in fuel meat shall be determined by using the fact that each fuel meat made of U-Al alloy contains all amount of U per fuel plate, and assuming that the rest is the volume of Al. Note that the detailed dimensions of the fuel assembly and the number of fuel plates loaded per fuel assembly (calculated as the maximum number of fuel plates) vary from core to core.

From the above, using the thermal neutron temperature T_n, the average activation cross section σ_{act} of Au foil for thermal neutrons is obtained using Eq. (6.11). In doing so, the non-$1/\upsilon$ correction factor $g(T_n)$ is obtained by interpolation using Table 6.1, and the values listed in Table 6.1 are used for the activation cross section $\sigma(\upsilon_p)$ of the Au foil corresponding to $\upsilon_p = 2200$ m s^{-1}.

(5) Perturbation factor f

Assuming that f is equal to f_2, determine it from Fig. 6.8 or Eq. (6.18) using the thickness t of the Au foil used and the macroscopic cross section $\sum_a$ obtained from the average activation cross section for thermal neutrons σ_{act}.

Thus, a complicated calculation is required to obtain the absolute value of thermal neutron flux from the radioactivity measurement of Au foil.

6.6 Discussions

Items (1) to (3) in Sect. 6.6.1 will be discussed as a summary of the experimental report. Section 6.6.2 will give the data processing and discuss the absolute measurement of thermal neutron flux by Au foil and determination of reactor power.

6.6.1 Experimental Reports

(1) Purpose of experiment,
(2) Experimental procedures,

(3) Experimental settings.

- Name of core, core layout and irradiation position of Au wire (refer to relevant documents),
- Operation number (Run No.) during irradiation, positions of control rods and safety rods, water level, core temperature, indicated values of linear power meter and γ-ray area monitor, start time, and end time of Au wire irradiation,
- Type of γ-ray detector used to measure radioactivity.

6.6.2 Determination of Thermal Neutron Flux and Reactor Power Using Gold Foils

(1) Measured data of radioactivity

For the results of the radioactivity measurements of Au foils, obtain the following items for each of the bare and Cd-covered Au foils:

- Locations of Au foils in the core,
- Start time of radioactivity measurement,
- Irradiation time,
- Waiting time,
- Measurement time,
- Counts and count rates for the γ-ray at 411.8 keV,
- Background counts and count rates,
- Net count rates of the γ-ray at 411.8 keV (γ-ray count rates—background count rates),
- Data on radioactivity measurements using standard sources,
- Detection efficiency of the γ-ray at 411.8 keV,
- Saturated activity in Eq. (5.15),
- Mass of Au foil samples,
- Reaction rate per mass in Eq. (5.16).

(2) Discussions

- Based on the results of the radioactivity measurements of Au foils, determine the absolute values of thermal neutron flux and reactor power at the irradiation location according to the item (3) in Sect. 6.7.2. Also include the Cd ratio obtained by the Au foils and the various other factors used in the correction.
- Using the results of the relative reaction rate distributions from the Au wires in directions x and z obtained in Chap. 5, evaluate quantitatively the relationship between the irradiation position and the center of the fuel region with respect to the reaction rates, and determine the absolute values of thermal neutron flux and reactor power at the core center (centers in directions x, y, and z).

6.7 Preliminary Exercises

6.7.1 Relationship Between Thermal Neutron Flux and Reactor Power

In a rectangular core, if the lengths of sides in directions *x*, *y*, and *z* are *a*, *b*, and *c* (cm), respectively, and the reflector savings are δ_x, δ_y, and δ_z (cm), the extrapolated lengths of each side can be expressed as $a^* = a + \delta_x, b^* = b + \delta_y$, and $c^* = c + \delta_z$ (cm). Assuming that the thermal neutron flux forms a cosine distribution in directions *x*, *y*, and *z*, if the thermal neutron flux at the center position of the core is ϕ_0 ($\text{cm}^{-2}\,\text{s}^{-1}$), the average thermal neutron flux in the core ϕ_{av} ($\text{cm}^{-2}\,\text{s}^{-1}$) can be expressed as follows:

$$\phi_{av} = \phi_0 \frac{8a^*b^*c^*}{\pi^3 abc} \sin\left(\frac{\pi a}{2a^*}\right) \sin\left(\frac{\pi b}{2b^*}\right) \sin\left(\frac{\pi c}{2c^*}\right). \tag{6.21}$$

The reactor power *P* (W) can be expressed as follows, using the macroscopic fission cross section of the core region $\sum_f$ and the energy released in one fission per unit time γ:

$$P = \gamma \sum\nolimits_f abc\phi_{av}. \tag{6.22}$$

Derive Eqs. (6.21) and (6.22) shown above.

6.7.2 Determination of Thermal Neutron Flux and Reactor Power Using Gold Foils

Assume that bare and Cd-covered Au foils are set at the center of a C45G0(5) core (HEU core in Ref. [4]) and irradiated, and that after the radiation dose in the core has sufficiently decreased, the Au foils are removed, and the γ-ray at 411.8 keV emitted from ^{198}Au is measured with the HPGe detector. The diameter and thickness of the Au foil are assumed to be 9.5 and 0.05 mm, respectively, and the thickness of the Cd cover is 0.8 mm. Also, assume that the detection efficiency of the HPGe detector for the γ-ray at 411.8 keV is 28.9%.

Assuming that the data obtained from measurements of the Au foils were as in items (1) and (2), resolve item (3) as follows:

(1) Data of Au foil irradiation

- Irradiation start time: 16:30,
- Irradiation end time: 17:00,
- Core temperature: 300.5 K,
- Indicated value of reactor power: 50 W.

Table 6.3 Data obtained from measurements of radioactivity of Au foils

Sample No	Measurement start time	Measurement time (s)	Counts
BG		300	316
1 (Bare)	11:30	60	31,521
2 (Cd-covered)	11:40	180	28,846
1 (Bare)	13:30	60	30,909
2 (Cd-covered)	13:40	180	28,216
BG		300	321

(2) Data obtained from measurements of radioactivity of Au foils (Table 6.3).

(3) Parameters to be obtained

- Absolute values of the saturated radioactivity of bare and Cd-covered Au foils,
- Cd ratio $(RR)_{\mathrm{Cd}}$,
- F_{Cd},
- Thermal neutron transmittance $(TR)_{\mathrm{Cd}}$,
- Neutron temperature T_n,
- Average activation cross section σ_{act} of Au foil for thermal neutrons,
- Perturbation factor f,
- Absolute value of the thermal neutron flux at the center position of the core ϕ_{th} using Eq. (6.21),
- Reactor power P (W) using Eq. (6.22).
- (The information could be provided later, including geometry and reactor constants of the experimental core used in Eqs. (6.21) and (6.22).)

6.8 Absolute Measurement of Activated Reaction Rates by Coincidence Counting

6.8.1 Principle of $4\pi\beta - \gamma$ Coincidence Counting

Focusing on the phenomenon of simultaneous emission of β-ray and γ-ray in Au decay, the count rates C_β, C_γ, and $C_{\beta\gamma}$ were measured for β-ray, γ-ray, and β-ray and γ-ray coincidences, respectively, to determine the absolute value of decay rate D of ^{198}Au generated in Au foil, and finally the absolute value of the saturated activity D_∞. In the $4\pi\beta - \gamma$ coincidence counting, a 4π gas flow counter is used as the β-ray detector and two large NaI(Tl) scintillation detectors surrounding it as the γ-ray detectors. If the detection efficiencies of β-ray and γ-ray detectors for the Au foil are ε_β and ε_γ, the count rates C_β, C_γ, and $C_{\beta\gamma}$ can be written as follows, respectively:

$$C_\beta = D\varepsilon_\beta \left\{ 1 + \frac{(1-\varepsilon_\beta)\varepsilon_c \alpha}{\varepsilon_\beta (1+\alpha)} \right\}, \tag{6.23}$$

$$C_\gamma = \frac{D\varepsilon_\gamma}{1+\alpha}, \tag{6.24}$$

$$C_{\beta\gamma} = \frac{D\varepsilon_\beta \varepsilon_\gamma}{1+\alpha}, \tag{6.25}$$

where α is the internal conversion factor (Sect. 6.2.1) of γ-ray at 411.8 keV and ε_c is the detection efficiency of the 4π counter for internal conversion electrons in the energy transition at 411.8 keV (^{198}Au).

In this case, the decay rate D is as follows:

$$D = \frac{C_\beta C_\gamma}{C_{\beta\gamma} \left\{ 1 + \frac{(1-\varepsilon_\beta)\varepsilon_c \alpha}{\varepsilon_\beta (1+\alpha)} \right\}}, \tag{6.26}$$

where the internal conversion factor α of the γ-ray at 411.8 keV is 0.041 (Ref. [3]). The value of ε_β is then obtained from the measurements of C_γ and $C_{\beta\gamma}$ as follows:

$$\varepsilon_\beta = \frac{C_{\beta\gamma}}{C_\gamma}. \tag{6.27}$$

The value of ε_c is the detection efficiency of the 4π counter for conversion electrons that can be emitted in the energy transition at 411.8 keV, which is mainly determined by the thickness of the Au foil. The results of ε_c are shown in Table 6.4.

Equation (6.26) was obtained by simplifying the ^{198}Au decay scheme in Fig. 6.1. In addition, the specific radioactivity in the Au foil was assumed to be constant, and ε_β and ε_γ at any position in the foil were also constant. For the latter requirement, the Au foil should be as small and thin as possible, which also reduces the amount of foil activation. However, if 4π counters with the highest geometric efficiency are employed for the β-ray detectors and large NaI(Tl) scintillators for the γ-ray detectors, the high statistical accuracy can be maintained by paying attention to the values of neutron flux density and foil mass.

Table 6.4 Detection efficiencies ε_c of 4π counter for internal conversion electrons from γ-ray transition at 411.8 keV

Thickness of Au foil (mg cm^{-2})	ε_c
6	0.975
31	0.880
49	0.820
110	0.610

Note the contribution from γ-ray detection of the 4π counter is ignored here

When using Eq. (6.26) to determine the absolute value of the decay rate, the following corrections should be made to the count rates C_β, C_γ, and $C_{\beta\gamma}$:

- Correction of background for C_β, C_γ, and $C_{\beta\gamma}$,
- Correction of signal decomposition time for C_β,
- Correction of accidental coincidence for $C_{\beta\gamma}$.

Assuming the decomposition time of the coincidence circuit is 2τ, the accidental coincidence C_{ch} is expressed as follows:

$$C_{ch} = 2\tau C_\beta C_\gamma. \tag{6.28}$$

Here, C_{ch} should be less than 10% of $C_{\beta\gamma}$. Therefore, the decay rate of the sample should be $D \leq 0.1 \times \left(\frac{1}{2\tau}\right)$, since the relationship $2\pi\varepsilon_\beta\varepsilon_\gamma D^2 \rightarrow 0.1\varepsilon_\beta\varepsilon_\gamma$ holds. As an example, for $\tau = 1\mu$s, D should be less than $0.1 \times \left(\frac{1}{2\times10^{-6}}\right) = 5 \times 10^4$ Bq ($\simeq 1.35\,\mu$ Ci).

6.8.2 Absolute Values by $4\pi\beta - \gamma$ Coincidence Counting

We make an attempt to determine the radioactivity of ^{198}Au by irradiated bare and Cd-covered Au foils using the $4\pi\beta - \gamma$ coincidence counting. The β and γ, and $\beta\gamma$ coincidence count rates are measured for each foil. The detector body of the $4\pi\beta - \gamma$ coincidence counting system is shown in Fig. 6.9. The β-ray detector is a square 4π counter with a Mylar source support film deposited by gold to provide conductivity in the center, and gas flow type proportional counters set at upper and lower parts. The counting gas (PR gas) flows through this counter. The γ-ray detector consists of two NaI(Tl) scintillation detectors by which the β-ray detectors are sandwiched. The output of each detector for β-ray and γ-ray is fed into a pre-amplifier and main amplifier, and logic pulses generated in a single-channel pulse height analyzer are then fed into a single and coincidence connecting circuits to obtain the count rates C_β, C_γ, and $C_{\beta\gamma}$.

Note that the following items should be taken into account when handling the β-ray detector:

- Carefully insert and remove the sample to avoid damage to the source support membrane.
- The applied voltage of the detector is set to zero during sample exchange.
- Avoid cutting the platinum wire electrode under the source support membrane.

The count rates of background $(C_\beta)_{\mathrm{BG}}$, $(C_\gamma)_{\mathrm{BG}}$, and $(C_{\beta\gamma})_{\mathrm{BG}}$ of the count rates C_β, C_γ, and $C_{\beta\gamma}$ are obtained before and after the start of the measurement. If all the count rates measured at time T_c after the waiting time T_w are $C_{\beta,total}$, $C_{\gamma,total}$, and $C_{\beta\gamma,total}$, respectively, the average count rates $C_{\beta,av}$, $C_{\gamma,av}$, and $C_{\beta\gamma,av}$ can be expressed as follows, respectively:

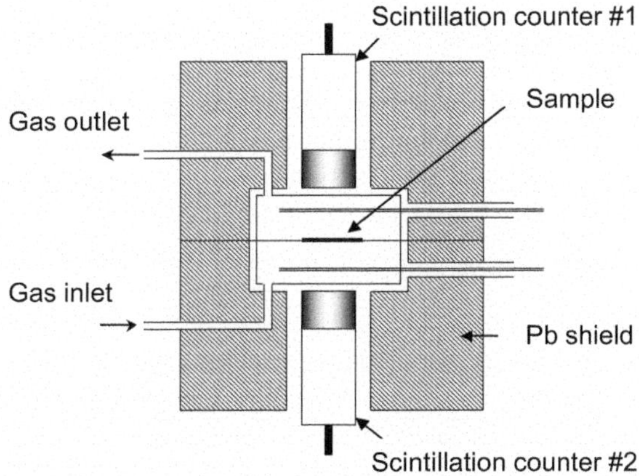

Fig. 6.9 Schematic view of $4\pi\beta - \gamma$ coincidence detector (Ref. [4])

$$C_{\beta,av} = \frac{\left(\frac{C_{\beta,total}}{T_c}\right)}{1 - C_{\beta,total}\tau_\beta} - (C_\beta)_{\mathrm{BG}}, \tag{6.29}$$

$$C_{\gamma,av} = \left(\frac{C_{\gamma,total}}{T_c}\right) - (C_\gamma)_{\mathrm{BG}}, \tag{6.30}$$

$$C_{\beta\gamma,av} = \left(\frac{C_{\beta\gamma,total}}{T_c}\right) - 2\tau\left(\frac{C_{\beta,total}}{T_c}\right)\left(\frac{C_{\gamma,total}}{T_c}\right), \tag{6.31}$$

where counting losses in β and $\beta\gamma$ measurement systems were neglected. The value of 2τ is the decomposition time of the coincidence circuit.

Substituting Eqs. (6.29) through (6.31) into Eq. (6.26) to obtain the average decay rate D and substituting it into $D_\infty = D/\left(1 - e^{-\lambda T_i}\right)e^{-\lambda T_w}$($\lambda$: decay constant; T_i: irradiation time; and T_w: waiting time), we can obtain the saturated activity D_∞, and thus the saturated activity $(D_\infty)_{\mathrm{bare}}$ and $(D_\infty)_{\mathrm{Cd}}$ for the irradiated bare and Cd-covered Au foils, respectively.

6.9 Overview of High-Purity Germanium Detector

The HPGe detector consists of a high-purity Ge single-crystal, a cold finger, a cryostat, a preamplifier, and a liquid nitrogen container (Dewar). The structure of HPGe detector is shown in Fig. 6.10 (well-type). Various shapes of cryostats and liquid nitrogen containers are also made according to the purpose of use. Usually, the γ-ray is injected through the tip of the cryostat, called the end cap, which is made of Al[7] about 0.5 mm in diameter. The end cap part of the cryostat has a narrow gap with

[7] For detectors in measuring γ-ray or X-ray with low energy less than a few tens of keV, the beryllium window detectors are used.

the Ge crystal and weak mechanical strength, so care must be taken not to shock the sample when handling it.

The block diagram of a standard measurement system is shown in Fig. 6.11. The configuration is almost the same as that of a well-type NaI(Tl) scintillation detector used to measure the reaction rate distribution by Au wire.

The precautions for handling the HPGe detector are shown as follows:

(1) Detector cooling

The HPGe detector can be stored under room temperature, but the detector should be completely cooled at the time of use. If a bias voltage is accidentally applied to the detector at room temperature with incomplete cooling, the Ge crystals, and the field effect transistor (FET) in the pre-amplifier will be damaged. The time required for cooling varies depending on the structure of the cryostat, but the most widely

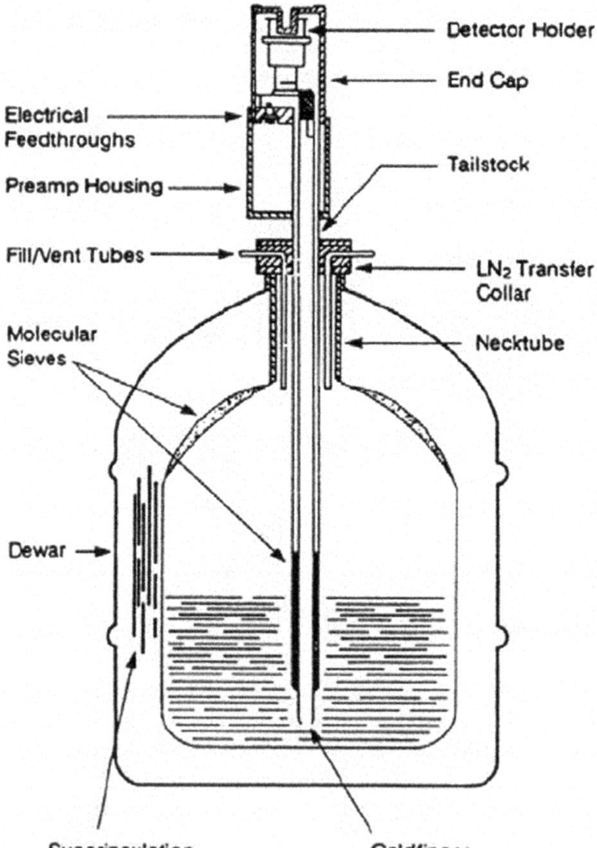

Fig. 6.10 Structure of HPGe detector (Ref. [4])

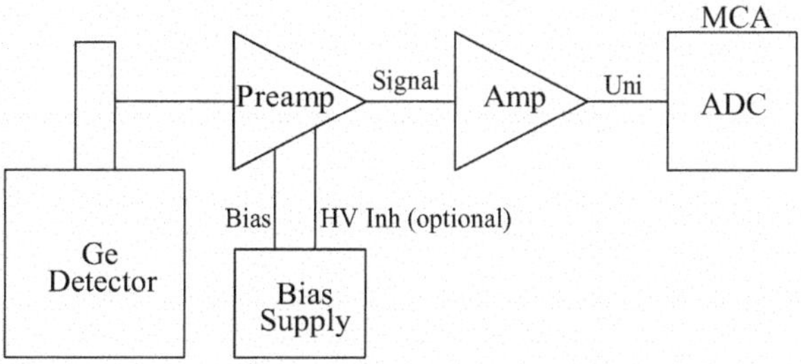

Fig. 6.11 Typical blocks of measurement system by HPGe detector (Ref. [4])

used I-type (and inverted L-type) cryostats take about 6–8 h. In addition, cooling for about 12 h is desirable for stable measurement.

(2) Liquid nitrogen replenishment

During the measurement, special care should be taken to prevent the temperature of the crystalline part from rising due to the lack of liquid nitrogen. In particular, small Dewars may require replenishment of liquid nitrogen during measurement.

(3) Re-cooling of detector

If the liquid nitrogen decreases and the temperature of the cryostat is higher than the liquid nitrogen temperature, the liquid nitrogen in the Dewar must be eliminated (or the cryostat removed from the Dewar), the temperature of the cryostat section must be completely brought to room temperature, allowed to stand for several hours, and then cooling must begin again.

(4) How to apply detector bias voltage

When applying a bias voltage to the HPGe detector, the voltage should be applied slowly at a ramp-up rate of 100 V s^{-1} or less. This slow voltage ramp-up is a common precaution when using the preamplifier that sets an FET in the first stage. Therefore, special care should be taken with HPGe detectors, whether ramp-up or down, since a sudden change in bias voltage can damage not only the pre-amplifier but also the Ge crystal.

(5) Use of protective devices

The following devices are recommended to protect the detector: from inadvertent bias voltage application or sudden power failure, an integrating circuit with a very large time constant, called a high-voltage buffer circuit, should be placed between the bias power supply and the detector. To prevent the crystal temperature from rising due to the lack of liquid nitrogen, it is recommended to use the bias power supply that can detect the level of the liquid nitrogen in the Dewar, and automatically turn off the bias power supply when the liquid nitrogen level decreases lower than a set value.

References

1. Knoll GF (2010) Radiation detection and measurements, 4th edn. John Wiley & Sons Inc., New Jersey
2. Firestone RB, Baglin CM, Chu SYF (eds) (1998) Table of isotopes, 8th edn. Wiley, New York
3. Gilmore G, Hemingway JD (1995) Practical gamma-ray spectrometry. John Wiley & Sons Ltd., Chichester, England
4. Misawa T, Unesaki H, Pyeon CH (2010) Nuclear reactor physics experiments. Kyoto University Press, Kyoto, Japan http://hdl.handle.net/2433/276400. Accessed 22 May 2024
5. Beckurts KH, Writs K (1964) Neutron physic. Springer-Verlag, Berlin, Germany
6. Jakeman D (1966) Physics of nuclear reactors. American Elsevier Pub. Com. Inc., New York
7. Lamarsh JR (1966) Introduction to nuclear reactor theory. Addison-Wesley Pub. Com. Inc., Massachusetts
8. Shibata K, Iwamoto O, Nakagawa T et al (2011) JENDL-4.0: a new library for nuclear science and engineering. J Nucl Sci Technol 48:1–30. https://doi.org/10.1080/18811248.2011.9711675
9. Helm FH (1963) Numerical determination of flux perturbation by foils. Nucl Sci Eng 16:235–238 https://doi.org/10.13182/NSE63-A26505

Chapter 7
Exponential Experiments with Natural Uranium

Abstract The purposes of the exponential experiments are to obtain reactor physics parameters for determining the criticality conditions of a nuclear reactor, and to confirm the feasibility of a newly proposed reactor concept. When conducting the experiments and modifying various methodologies and reactor constants used in the calculations, a target reactor can be brought closer to realization. The final goal of the exponential experiments is to realize a subcritical steady-state reactor with an external neutron source, and to determine neutronics characteristics of the reactor by measuring spatial neutron flux distributions in a subcritical system. In this chapter, an experimental system is introduced for conducting exponential experiments using natural uranium and polyethylene reflector at KUCA. Moreover, theoretical preparation and experimental procedures are described to obtain the reactor physics parameters, including the multiplication factor, spatial decay constant, and material buckling using reaction rate distributions measured by neutron detectors.

Keywords Exponential experiments · Natural uranium · Subcritical system · Multiplication factor · Spatial decay constant · Material buckling

7.1 Background

Most people who have studied nuclear energy know that the world's first exponential experiments using natural uranium (NU) were conducted in the United States, 1942, and were promoted mainly by Dr. Enrico Fermi, the Nobel Prize winner in physics. A very large reactor (graphite pile) was built to generate fission chain reactions using NU fuel and graphite reflector, and the reactor was named Chicago Pile No. 1 (CP-1; Ref. [1]). The purpose of the exponential experiments performed at CP-1 was to obtain the infinite multiplication factor, material buckling, and migration area, which are important parameters for determining the criticality conditions of a nuclear reactor. At a time when computer technology was not as advanced as it is today, through the CP-1 exponential experiments the attempt was made to obtain

C. H. Pyeon et al., *Reactor Laboratory Experiments at Kyoto University Critical Assembly*, https://doi.org/10.1007/978-981-97-8070-9_7

the design information necessary to achieve a critical reactor without spending large costs on conducting the experiments.

In general, the main purpose of conducting exponential experiments is to confirm the feasibility of a newly proposed reactor concept. Then, through the process of examining the factors causing differences between the design and the proposed reactor, conducting the experiments, and modifying various methodologies and reactor constants used in the calculations, the target reactor can be brought closer to realization.

The reactor in which the exponential experiments are carried out has a core in which the core parameters can be easily changed, and the core configuration is the same as that of the target reactor. The size of the core is then small in comparison with that of the target core. This attempt is like carrying out current mockup experiments with the use of a critical assembly, and it is evident that the exponential experiments by Dr. Enrico Fermi were already positioned as mockup experiments with the target reactor core in mind at that time. It is noteworthy here that the core in which the exponential experiments were conducted was a subcritical system. To maintain fission chain reactions with a steady state in a subcritical system, it is important to continuously supply neutrons from an external neutron source. Finally, the goal of the exponential experiments is to realize a subcritical steady-state reactor with an external neutron source and to determine neutronics characteristics of the reactor by measuring spatial neutron flux distributions in a subcritical system. In the next section, we will touch on the characteristics of the exponential experiments using an external neutron source.

7.2 Overview

Exponential experiments using an external neutron source can utilize neutrons produced by an accelerator or a radioactive element neutron source. When introducing the external neutron source into a subcritical system, it is easy to infer that steady-state neutron flux distributions in the subcritical system decrease exponentially with distance from the neutron source.

The characteristics of the subcritical system in exponential experiments are possible to reduce the risks associated with radiation exposure and allow measurements to be made in an environment with very low levels of radioactivity. The exponential experiments result in the measurement advantage of being able to detect neutrons using a simple measurement system. Reactor physics parameters that can be measured by exponential experiments include the spatial decay constant, diffusion length and cadmium (Cd) ratio. As a result, with the use of these, we can easily deduce the buckling, multiplication factor, the Fermi age, extrapolation distance, thermal neutron utilization factor, and reflector saving, which can cover a significant portion of reactor physics experiments carried out in a critical system.

Since the size of the core with the subcritical system used in exponential experiment is small, it is necessary to consider the effect of neutrons leaking outside the system. In other words, it is necessary to correct the neutron flux at the boundary between the fuel and the reflector or at the outer edge of the core for the measurement results. In addition, since the core is a subcritical system, the neutron flux in the core is low, and a high-count rate cannot be expected in measurements of neutrons.

7.3 Spatial Decay Constant

7.3.1 Theoretical Background

As a system for conducting exponential experiments, a rectangular reactor consisting of fuel is assumed to be given in a homogeneous system at a critical state without an external neutron source. As shown in Fig. 7.1, consider a bare rectangular reactor with lengths *a*, *b*, and *c*, including extrapolation distances, in directions *x*, *y*, and *z*, respectively. The origin of the coordinates is the center of the rectangular reactor, in which the Laplace operator ∇^2 in the rectangular coordinates (x, y, z) is expressed as follows:

$$\nabla^2 = \frac{\partial^2}{\partial x^2} + \frac{\partial^2}{\partial y^2} + \frac{\partial^2}{\partial z^2}. \tag{7.1}$$

Let the thermal neutron flux of the system in perpendicular coordinates $\phi(x, y, z)$ and the curvature determined from the geometry of the system (geometric buckling: B_g^2), the equation is assumed to be given as follows:

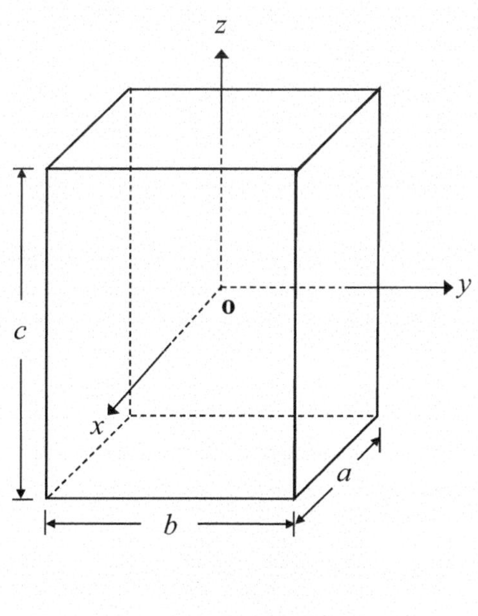

Fig. 7.1 Rectangular reactor (the origin is the center of the rectangle)

$$\left(\frac{\partial^2}{\partial x^2}+\frac{\partial^2}{\partial y^2}+\frac{\partial^2}{\partial z^2}\right)\phi(x,y,z)+B_g^2\phi(x,y,z)=0, \tag{7.2}$$

where the boundary condition $\phi(x, y, z) = 0$ is used on the extrapolated boundary of the system $x = \pm\frac{a}{2}$, $y = \pm\frac{b}{2}$, or $z = \pm\frac{c}{2}$ in Eq. (7.2).

Assuming that the thermal neutron flux $\phi(x, y, z)$ is separable with respect to the variables x, y, and z, $\phi(x, y, z)$ in Eq. (7.2) is transformed as follows:

$$\phi(x,y,z)=X(x)Y(y)Z(z), \tag{7.3}$$

where $X(x)$ is a function of variable x, and $Y(y)$ and $Z(z)$ are likewise functions of variables y and z, respectively.

Substituting Eq. (7.3) into Eq. (7.2), the following equation is obtained:

$$\frac{1}{X}\frac{d^2}{dx^2}X+\frac{1}{Y}\frac{d^2}{dy^2}Y+\frac{1}{Z}\frac{d^2}{dz^2}Z+B_g^2=0. \tag{7.4}$$

Since each term on the left-hand side of Eq. (7.4) is a function on its respective variable, each term can be made equal to a certain constant as follows:

$$\frac{1}{X}\frac{d^2}{dx^2}X=-\alpha^2, \tag{7.5}$$

$$\frac{1}{Y}\frac{d^2}{dy^2}Y=-\beta^2, \tag{7.6}$$

$$\frac{1}{Z}\frac{d^2}{dz^2}Z=-\gamma^2, \tag{7.7}$$

where α^2, β^2, and γ^2 are positive real numbers.

Furthermore, substituting Eqs. (7.5) through (7.7) into Eq. (7.4), the geometric buckling B_g^2 can be expressed as follows:

$$\alpha^2+\beta^2+\gamma^2=B_g^2. \tag{7.8}$$

From the symmetry of the thermal neutron flux distribution in the system and the boundary condition that the thermal neutron flux on the extrapolated boundary is zero, for example, the solution of Eq. (7.5) is as follows:

$$X(x)=A\cos(\alpha x), \tag{7.9}$$

where A is an arbitrary constant. Also, using the boundary condition that the thermal neutron flux is zero at $x = \frac{a}{2}$, the minimum value of α is $\alpha = \frac{\pi}{a}$ and Eq. (7.9) can be expressed as follows:

$$X(x) = A\cos\left(\frac{\pi}{a}x\right). \tag{7.10}$$

Applying the same conditions to $Y(y)$ and $Z(z)$ as to $X(x)$, the minimum values of β and γ are $\beta = \frac{\pi}{b}$ and $\gamma = \frac{\pi}{c}$, respectively, and the geometric buckling of Eq. (7.8) can be expressed as follows:

$$B_g^2 = \left(\frac{\pi}{a}\right)^2 + \left(\frac{\pi}{b}\right)^2 + \left(\frac{\pi}{c}\right)^2. \tag{7.11}$$

The thermal neutron flux $\phi(x, y, z)$ in Eq. (7.3) can be expressed using an arbitrary constant A' as follows:

$$\phi(x, y, z) = A'\cos\left(\frac{\pi}{a}x\right)\cos\left(\frac{\pi}{b}y\right)\cos\left(\frac{\pi}{c}z\right). \tag{7.12}$$

7.3.2 Derivation of Spatial Decay Constant

To conduct exponential experiments, it is necessary that lattices of the actual reactor and the system in the exponential experiments are the same. Also, the size of the core in the exponential experiments needs to be quite small in directions x and y: a subcritical assembly. Although no continuous chain reaction of fission can occur in a subcritical assembly, it is possible to achieve a steady state by using an external neutron source. As shown in Fig. 7.2, let us assume that the origin of the rectangular body is the center of the x–y plane of $z = 0$ and that the neutron source is placed at the location of the origin of the rectangular body.

How can the spatial distribution of thermal neutron flux in the core be expressed in the presence of an external neutron source? If an external neutron source is not introduced in the reactor, the equation can be obtained by using $\phi(x, y, z)$ as follows:

$$\left(\frac{\partial^2}{\partial x^2} + \frac{\partial^2}{\partial y^2} + \frac{\partial^2}{\partial z^2}\right)\phi(x, y, z) + B_m^2\phi(x, y, z) = 0, \tag{7.13}$$

where B_m^2 is the material buckling determined from a unit fuel cell. Here, the thermal neutron fluxes at the boundary including extrapolated distance in directions x and y, and at $z = c$ including extrapolated distance in the direction z, are assumed to be zero.

Assuming, as in Sect. 7.3.1, that the thermal neutron flux $\phi(x, y, z)$ is separable with respect to the variables x, y, and z, the following equation can be obtained using Eq. (7.13):

$$\frac{1}{X}\frac{d^2}{dx^2}X + \frac{1}{Y}\frac{d^2}{dy^2}Y + \frac{1}{Z}\frac{d^2}{dz^2}Z + B_m^2 = 0. \tag{7.14}$$

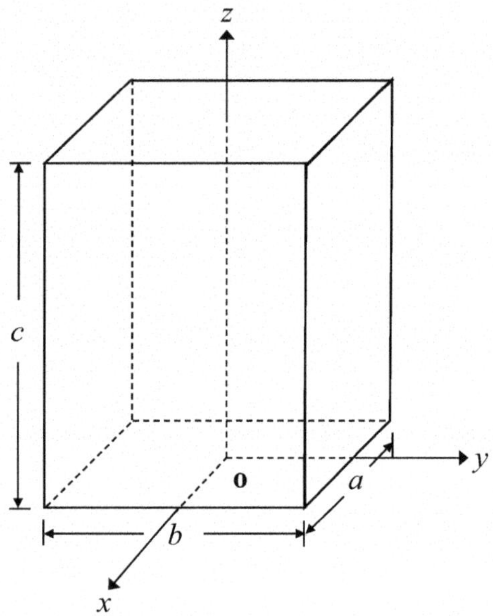

Fig. 7.2 Rectangular reactor (the origin is the center of the bottom surface of the rectangle)

Based on the boundary conditions in the direction z, the equation can be obtained as follows (see Ref. [2] in detail):

$$\frac{1}{Z}\frac{d^2}{dz^2}Z = \mu^2. \tag{7.15}$$

Using Eqs. (7.5), (7.6), and (7.15), Eq. (7.14) can be expressed as follows:

$$\alpha^2 + \beta^2 - \mu^2 = B_m^2, \tag{7.16}$$

where α^2, β^2, and μ^2 are positive real numbers. In this case, $X(x)$ and $Y(y)$ can be expressed as trigonometric functions as shown in Sect. 7.3.1. Since μ^2 is a positive real number, $Z(z)$ in Eq. (7.15) is expressed as hyperbolic functions as follows:

$$Z(z) = C_1 \cosh(\mu z) + C_2 \sinh(\mu z), \tag{7.17}$$

where C_1 and C_2 are arbitrary constants. For $z = c$, applying the condition that the thermal neutron flux is zero, the relationship between C_1 and C_2 is obtained as follows:

$$C_2 = -C_1 \frac{\cosh(\mu c)}{\sinh(\mu c)}. \tag{7.18}$$

Substituting Eq. (7.18) into Eq. (7.17), the following equation is obtained:

$$
\begin{aligned}
Z(z) &= C_1 \cosh(\mu z) - C_1 \frac{\cosh(\mu c)}{\sinh(\mu c)} \sinh(\mu z) \\
&= \frac{C_1}{\sinh(\mu c)} (\sinh(\mu c) \cosh(\mu z) - \cosh(\mu c) \sinh(\mu z)) \\
&= C_3 \sinh \{\mu(c - z)\},
\end{aligned}
\tag{7.19}
$$

where C_3 is an arbitrary constant and contains the constant C_1. By transforming Eq. (7.19) as an exponential difference, the following equation is obtained:

$$
Z(z) = \frac{C_3}{2} \left(e^{\mu(c-z)} - e^{-\mu(c-z)}\right) = C\, e^{-\mu z} \left(1 - e^{-2\mu(c-z)}\right), \tag{7.20}
$$

where C is an arbitrary constant and contains the constant C_3. When the rectangle length c is sufficiently large, Eq. (7.20) can be expressed approximately as follows:

$$
Z(z) \approx C\, e^{-\mu z}. \tag{7.21}
$$

Here, Eq. (7.21) shows that the thermal neutron flux of the system is decreased exponentially along an arbitrary straight line parallel to the direction z. In other words, Eq. (7.21) shows that the distance over which the thermal neutron flux decreases to approximately $1/e$ times is the value of $1/\mu$. The value of μ shown in Eq. (7.21) is then termed the spatial decay constant.

From the above, the thermal neutron flux $\phi(x, y, z)$ in the rectangular body shown in Fig. 7.2 can be expressed using an arbitrary constant A'' as follows:

$$
\phi(x, y, z) = A'' \cos\left(\frac{\pi}{a} x\right) \cos\left(\frac{\pi}{b} y\right) e^{-\mu z}. \tag{7.22}
$$

7.3.3 Discussion of Spatial Decay Constant

To exactly measure the spatial decay constant, the moderation length of moderator is discussed in this section. Let us assume that the neutron source is set at the origin of the rectangular body (Fig. 7.2) or in the negative direction from the origin, $z \leq 0$. The exponential experiments assume that almost all the neutrons from the external neutron source are moderated to thermal neutrons by the moderator (graphite, light water, heavy water, etc.). The assumption stands that neutrons from the external neutron source are moderated to thermal neutrons at least several times (about 2–3 times) of the slowing-down length farther from the plane of $z \leq 0$. Here, the slowing-down length $\sqrt{\tau}$, expressed as the square root of the Fermi age τ (Ref. [2]), is defined as the quantity that determines the rate at which neutrons leak out of a thermal neutron reactor of limited size during moderation. The values of the Fermi age and slowing-down length for the moderator are shown in Table 7.1.

Table 7.1 Fermi age and slowing-down length of moderators (Ref. [3])

Moderator	Fermi age τ (cm^2)	Slowing-down length $\sqrt{\tau}$ (cm)
Graphite (carbon: C)	350	18.7
Polyethylene* ($(C_2H_4)_n$)	107	10.3
Light water (H_2O)	33	5.7
Heavy water (D_2O)	120	11.0
Beryllium (Be)	98	9.9

*SRAC2006 [4] with JENDL-4.0 [5]

For example, it can be assumed that the neutrons from the external neutron source are sufficiently moderated to thermal neutrons at about three times of slowing-down length from the external neutron source. The way to confirm the estimation experimentally is to measure the Cd ratio that is the spectrum index of neutrons, directing toward $z > 0$. If the measurement result shows that the Cd ratio is constant with respect to $z > 0$, the assumption of exponential experiments that neutrons from the external neutron source are sufficiently moderated to thermal neutrons could be then satisfied.

7.3.4 *Diffusion Length*

Let us consider the following case, in which the system is somewhat special. Suppose that in an infinite plate system there is a non-multiplying homogeneous reactor. If the core is in a steady state and the rate of change of neutron density with time is zero, the behavior of neutrons can be expressed using neutron diffusion equation with one-energy group as follows:

$$-D\frac{d^2}{dz^2}\phi(z) + \sum\nolimits_{\mathrm{a}} \phi(z) = S, \tag{7.23}$$

where ϕ indicates the neutron flux, D the diffusion coefficient, $\sum_{\mathrm{a}}$ the macroscopic absorption cross section, and S the neutron source term. To solve Eq. (7.23), the condition of $S = 0$ is applied to regions other than where the external neutron source is set. The following equation is then obtained:

$$-D\frac{d^2}{dz^2}\phi(z) + \sum\nolimits_{\mathrm{a}} \phi(z) = 0. \tag{7.24}$$

Furthermore, Eq. (7.24) is transformed as follows:

$$-\frac{d^2}{dz^2}\phi(z) + \frac{\sum_{\mathrm{a}}}{D}\phi(z) = 0. \tag{7.25}$$

Since $\sum_a$ and D have an inverse value of length and length itself, respectively, the value of $L = \sqrt{D/\sum_a}$ has the dimension of a length, and L is then called the diffusion length of neutrons in the medium. The one-dimensional (direction z) neutron flux $\phi(z)$ defined by Eq. (7.25) can then be expressed using $L = \sqrt{D/\sum_a}$ as follows:

$$\phi(z) = \phi_0 \, e^{-\frac{\sum_a}{D} z} = \phi_0 \, e^{-\frac{1}{L} z}, \tag{7.26}$$

where ϕ_0 is an arbitrary constant.

Let us consider again the special case in which Eq. (7.26) is derived. If it is possible to create a thermal neutron source on an infinitely wide plane in an infinitely large medium, the diffusion length can be determined from Eq. (7.26) by measuring the change in neutron flux with respect to the length from the external neutron source, as follows:

$$\ln \phi(z) = \ln \phi_0 - \frac{1}{L} z \quad \Leftrightarrow \quad \frac{d}{dz} \ln \phi(z) = -\frac{1}{L}. \tag{7.27}$$

If $\ln \phi(z)$ in Eq. (7.27) can be obtained experimentally, i.e., by plotting the value of $\ln \phi(z)$ for length z, a straight line with slope, the value of – $(1/L)$, can be obtained. Since the actual experiment uses an external neutron source of finite size in a medium of finite size, it is not possible to analyze the experimental results by simply applying the equation for a flat source in an infinite medium as in Eq. (7.27). In an infinite medium, there is no net loss of neutrons leaking from the system, whereas in a finite medium in directions x and y, there is the restriction that the loss of neutrons leaking from the system should be taken into account.

Although the diffusion length varies depending on the type of moderator used in the experimental system, the thermal neutron diffusion length, macroscopic absorption cross sections, and diffusion coefficients for various moderators should be provided in a better insight into the diffusion length obtained from the experiments, as shown in Table 7.2.

Table 7.2 Diffusion parameters of moderators for thermal neutrons (one-energy group in Ref. [3])

Moderator	Density (g cm^{-3})	L (cm)	$\sum_a$ (cm^{-1})	D (cm)
C	1.62	50.1	0.00036	0.903
$(C_2H_4)_n$*	0.95	2.79	0.019	0.148
H_2O	1.00	2.89	0.017	0.142
D_2O	1.10	100.0	0.00008	0.80
Be	1.84	23.2	0.0013	0.70

*SRAC2006 with JENDL-4.0

7.3.5 Multiplication Factor

When considering the criticality of a reactor, we can imagine that the geometrical buckling B_g^2 is a small value for a large reactor, from the relationship between the size of the reactor and the buckling determined by the geometry of the core.

For a bare and homogeneous reactor in a critical state ($k_{\text{eff}} = 1$), the criticality equation (Eq. (2.14) in Chap. 2) can be expressed as follows:

$$k_{\text{eff}} = \frac{k_\infty\, e^{-B_g^2\,\tau}}{1 + L^2\, B_g^2} = 1, \tag{7.28}$$

where k_{eff} is the effective multiplication factor of the system, k_∞ the infinite multiplication factor, and τ, L^2 and B_g^2 are the parameters defined in Sects. 7.3.2 through 7.3.4. Here, when the reactor is large, k_∞ in the critical state is considered slightly larger than unity. If the value of $e^{-B_g^2\tau}$ in Eq. (7.28) is expanded in series and the second term is ignored, the following approximation is obtained for $e^{-B^2\tau}$:

$$e^{-B_g^2\,\tau} \approx 1 - B_g^2\,\tau \approx \left(1 + B_g^2\,\tau\right)^{-1}. \tag{7.29}$$

Applying the approximation of Eqs. (7.29) to (7.28), the critical equation can be approximated as follows:

$$\frac{k_\infty}{\left(1 + L^2\, B_g^2\right)\left(1 + B_g^2\,\tau\right)} \approx 1. \tag{7.30}$$

Since the value of B_g^2 is very small for large reactors, neglecting the second-order term with respect to B_g^2 in Eq. (7.30), the following approximation is obtained:

$$\frac{k_\infty}{1 + B_g^2(L^2 + \tau)} \approx 1, \tag{7.31}$$

where k_∞ is expressed by using the four-factor formula as follows:

$$k_\infty = \eta\,\varepsilon\, p f, \tag{7.32}$$

where η indicates the reproduction factor (the number of neutrons produced by fission per neutron absorbed by the fuel), ε the fast fission factor (the ratio of the number of fission neutrons due to thermal fission to the number of total fission neutrons), p the resonance escape probability, and f the thermal utilization factor.

When k_∞ is given by the four-factor formula, and L^2 and τ are obtained by experiments, the geometrical buckling can be approximated by using Eq. (7.31) as follows:

$$B_g^2 \approx \frac{k_\infty - 1}{L^2 + \tau} = \frac{k_\infty - 1}{M^2}, \tag{7.33}$$

where M^2 is generally defined as the migration area.

For example, for a reactor constituting of an NU fuel and graphite reflector, η is said to be 1.3, ε about 1.03, p and f both less than unity (Ref. [2]), and k_∞ is slightly greater than unity. This suggests that, to make an NU-fueled reactor in a critical state, the fraction of neutrons leaking from the system needs to be reduced, and as a result, the reactor could be inevitably larger. Furthermore, when the value of k_∞ is slightly larger than unity and is substituted into Eq. (7.33) to calculate the geometrical buckling, the value of geometrical buckling is very small: the hand calculation makes it easy to infer that the size of the reactor could be large.

As indicated above, the exponential experiments are concerned with measuring the spatial decay of neutron flux distribution, diffusion length, and neutron multiplication in a subcritical system. However, as mentioned in Sect. 7.1, the main purpose of the experiments is to confirm the feasibility of the newly proposed reactor concept and to provide more concrete design information necessary for a reactor designed for larger scale operation. Then, the goal is to realize a reactor that will always be critical, and the noteworthy is a series of the constraint that the system of exponential experiments is in a small subcritical reactor. It is also necessary to always keep in mind that an important goal of the exponential experiments is to extrapolate and extract the parameters assumed in the larger core from the data obtained in the experiments using the smaller core.

7.4 Exponential Experiments

7.4.1 Objectives

In experiments using a subcritical system, an external neutron source is inserted into the system to stabilize the system in a steady state, and the behavior of thermal neutrons is then observed. Although the presence of a neutron source in the system is different from that in a reactor with a critical system, measurement instrumentation used in the exponential experiments is the same as those used in the approach-to-criticality experiment (Chap. 3). Therefore, basic concepts and measurement methods are significantly the same as those of the approach-to-criticality experiment. Here, the objectives of conducting exponential experiments using a subcritical system are considered as follows:

- To measure the inverse multiplication for the subcritical system in the same way as the approach-to-criticality experiment and to confirm if we can estimate the parameters corresponding to the critical mass of the system.

- To confirm experimentally that neutrons in the system are multiplied by maintaining a steady-state subcritical system with an external neutron source and by increasing the number of fuel assemblies in the subcritical state.
- To measure, in a subcritical system using an external neutron source, reaction rate distributions corresponding to thermal neutron flux distributions in the system by using neutron detectors and to confirm these shapes.
- To deduce various parameters related to the subcritical multiplication system estimated from the experimental results, using reaction rate distributions obtained from neutron detectors.
- To confirm whether the main objective and goal of the experiments can be fulfilled as required by the exponential experiments.

7.4.2 Experimental Settings

To set the experimental system, a fuel assembly (Fig. 7.3) was made of 70 fuel cells consisting of an NU plate (2 × 2 × 1/24 in.; in. = 25.4 mm) and a polyethylene moderator plate (p; 2 × 2 × 1/4 in.). As described in Sect. 7.3.3, since the slowing-down length of the system composed of NU and polyethylene was about 300 mm, the effective length of the fuel region, about 520 mm, is considered sufficient as a prerequisite for conducting the exponential experiments, sandwiching between polyethylene reflectors (PE) about 500 and 470 mm on left and right sides, respectively.

Although it is different from the example of a fuel cell made of NU and polyethylene plates, Fig. 7.4 shows a photo of a fuel cell made of highly-enriched uranium (HEU) and polyethylene plates for reference.

As shown in Figs. 7.5 and 7.6, 24 fuel assemblies in Fig. 7.3 are arranged in a square, with polyethylene reflectors surrounding the fuel region. A side view of the fuel system is shown in Fig. 7.7. At the center location of the fuel region (5, E in Fig. 7.5), a polyethylene rod with 1/2 in. (1/2″) diameter hole is provided to allow insertion of an external neutron source (californium-252: ^{252}Cf). In addition, two

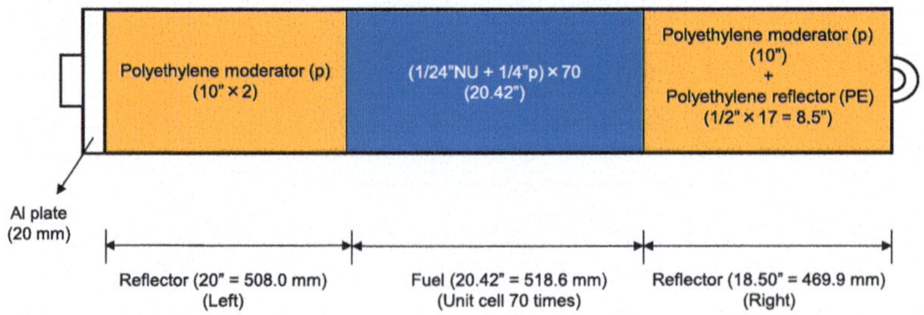

Fig. 7.3 Side view of fuel assembly in exponential experiments

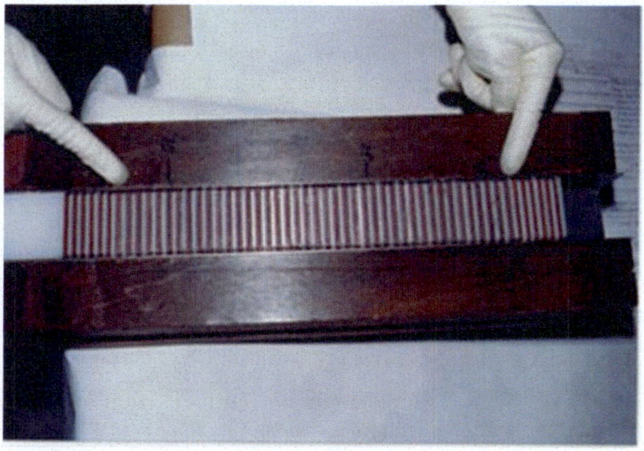

Fig. 7.4 Fuel region consisting of HEU and polyethylene plates.

polyethylene rods with 1″ diameter hole are provided in the reflector area to allow insertion of helium-3 (^{3}He) neutron detectors at (2, G) and (5, J) (both in Fig. 7.5).

As shown in Fig. 7.8, the origin of the coordinate system in the experimental system is the position where the ^{252}Cf neutron source is installed (5, E; Fig. 7.5). In the experiment, the ^{252}Cf neutron source is installed through an insertion hole in a 1/2″ perforated polyethylene reflector.

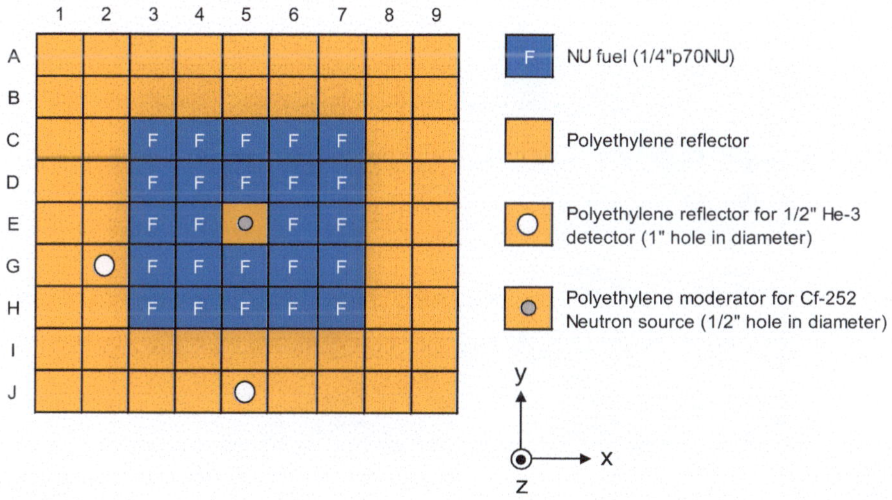

Fig. 7.5 Top view of experimental system

Fig. 7.6 Photo of top view of experimental system. © KURNS. All rights reserved

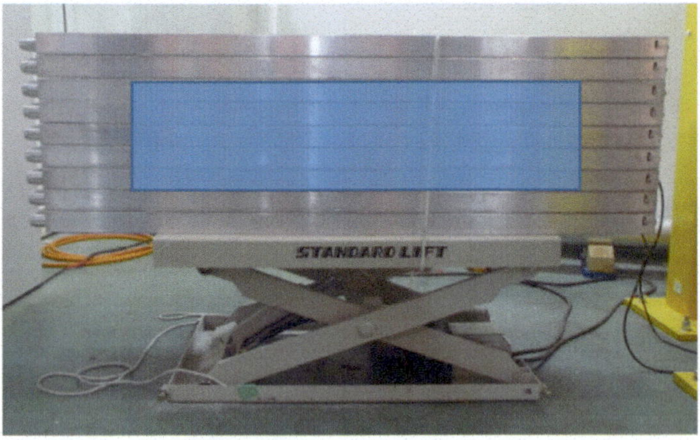

Fig. 7.7 Photo of side view of experimental system (blue center area is the fuel region). © KURNS. All rights reserved

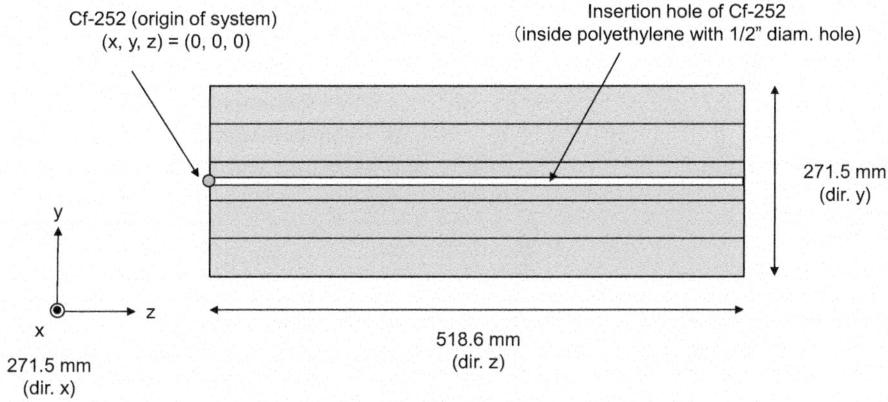

Fig. 7.8 Location of ^{252}Cf in experimental system

7.5 Procedures of Experiments

7.5.1 *Inverse Multiplication of Neutrons*

The neutron multiplication and its inverse value are obtained from the values counted by the neutron detector. If we follow the principle faithfully, the first state of the experiment should be that there is no fuel in the system at all, but for the convenience of the experiment, the first state should be that the system is loaded with fuel to some extent (hereinafter termed the initial state).

Let $k_{s,0}$ and $k_{s,i}$ the values of neutron multiplication for the initial system (with subscript 0) and the system to which fuel is added from the initial state (with subscript i), respectively (with subscript s for the rate of change, partly because the neutron source is inserted beforehand). Since the number of neutrons produced by fission reactions corresponds to the neutron fluxes $\phi_{s,0}$ and $\phi_{s,i}$ obtained by the neutron source, as in Chap. 3, $\phi_{s,0}$ and $\phi_{s,i}$ are approximately obtained as follows, respectively:

$$\phi_{s,0} \approx \frac{S}{1 - k_{s,0}}, \tag{7.34}$$

$$\phi_{s,i} \approx \frac{S}{1 - k_{s,i}}, \tag{7.35}$$

where S is the number of neutrons produced by the ^{252}Cf neutron source.

The ratio of the neutron flux when changing from the initial state 0 to state i, $\phi_{s,i}/\phi_{s,0}$, can be expressed using Eqs. (7.34) and (7.35) as follows:

$$\frac{\phi_{s,i}}{\phi_{s,0}} \approx \frac{1 - k_{s,0}}{1 - k_{s,i}}. \tag{7.36}$$

Since the initial state is the same for any state *i*, i.e., the numerator $1 - k_{s,0}$ on the right-hand side of Eq. (7.36) can be regarded as a constant value, and if $1 - k_{s,0}$ is the standardization factor, Eq. (7.36) can be approximated as follows:

$$\frac{\phi_{s,i}}{\phi_{s,0}} \approx \frac{1}{1 - k_{s,i}}. \tag{7.37}$$

The inverse value of Eq. (7.37) can be then expressed as follows:

$$\frac{\phi_{s,0}}{\phi_{s,i}} \approx 1 - k_{s,i}. \tag{7.38}$$

Equation (7.38) has the same form as the inverse of the multiplication M (inverse multiplication)$1/M \approx 1 - k_s(k_s$: multiplication factor in subcritical systems) in Chap. 3, and the method for finding the inverse multiplication $1/M$ can be directly applied when the inverse $\phi_{s,0}/\phi_{s,i}$ of the ratio of neutron flux in Eq. (7.36) is found (see Chap. 3).

Let us assume that the response per unit time (count rate) obtained from the neutron detector is approximately proportional to the neutron flux. Taking C_0 and C_i as the counts obtained in the initial system (subscript 0) and the state i changed by the addition of fuel, respectively, the ratio of count rates C_0/C_i can be expressed from Eq. (7.38) as follows:

$$\frac{C_0}{C_i} \approx \frac{\phi_{k,0}}{\phi_{k,i}} \approx 1 - k_{s,i}\left(= \frac{1}{M_i}\right). \tag{7.39}$$

The value of C_0/C_i obtained in Eq. (7.39) is called the inverse count rate (inverse multiplication in state *i*: $1/M_i$). In other words, experimentally, the inverse count rate in Eq. (7.39) is nothing but the inverse multiplication itself.

In the exponential experiments, the multiplication of neutrons in the system is achieved by increasing the number of fuel assemblies, as shown in Fig. 7.9. The patterns of increasing the number of fuel assemblies are shown in Fig. 7.9.

7.5.2 Reaction Rate Distribution by Neutron Detector

Measurement of the reaction rate distribution with a neutron detector is performed using the ^{3}He detector. The ^{3}He detector is a counting tube for detecting thermal neutrons. A metal tube with a core wire is filled with ^{3}He gas, and a high voltage is applied between the core wire and the tube wall. When a thermal neutron passes through, the $^3\mathrm{He(n,p)}^3\mathrm{H}$ reaction occurs, and the protons (p) and ^{3}H (tritium)

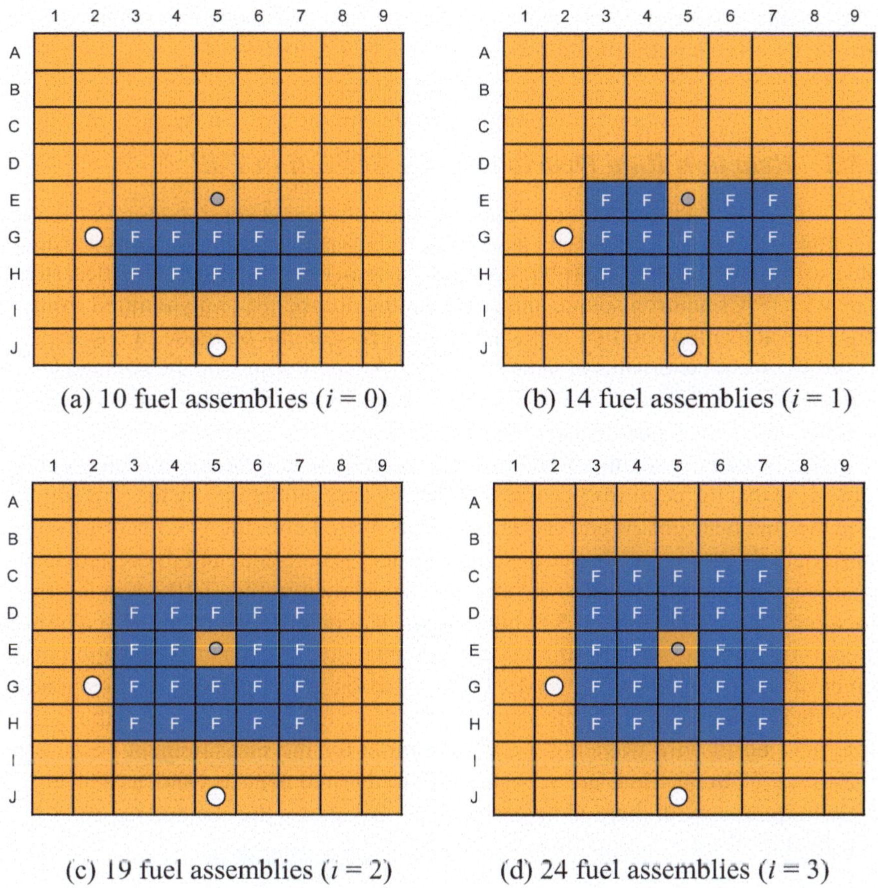

Fig. 7.9 Patterns of increase in number of fuel assemblies in inverse multiplication of neutrons

produced are set in motion by 765 keV of nuclear reactions, ionizing the gas. The ions cause an electrical discharge between the core wire and the tube wall, and the neutron passages can be counted as pulses. The ^{3}He detectors are set at positions (2, G) and (5, J) of the system, as shown in Fig. 7.9. Also, a neutron source (^{252}Cf) is fixed at the origin ($z = 0$) in direction z at (5, E).

The method of measuring the reaction rate distribution with a neutron detector involves measuring neutron counts while changing the neutron detector to a positive position in the direction z ($z = 0$, 100, 200, 300…mm) for the state i of the system (increase in fuel assemblies). Using this method, the reaction rate distribution in the direction z can be measured simply by changing the detector position, and the measurement time can be changed according to the values of neutron counts. On the other hand, the neutron flux distribution may be distorted (termed perturbation) by moving the detector, and several detector positions have different effects on the neutron flux distortion. Note that this reaction rate distribution is obtained when measuring inverse multiplication in state i ($i = 0, 1, 2$ and 3).

7.5.3 Reaction Rate Distribution by Activation Foil

The reaction rate distribution by activation foils is obtained by measuring reaction rates, using experimental data obtained from neutron irradiation in a subcritical steady state with ^{252}Cf neutron source and counting the number of γ-ray emitted from the foils. The activation foil that is the activation detector can be made of any material according to characteristics of neutron energy. Also, by adjusting the size and length of the activation foil so as not to distort the thermal neutron flux distribution in the system, the reaction rate distribution can be measured at any position.

In the system consisting of NU fuel and polyethylene reflector, it is easy enough to imagine that the neutrons generated from ^{252}Cf are moderated by the polyethylene to be thermal neutrons. When enough time has passed after the ^{252}Cf neutron source is inserted, the behavior of neutrons in the system is constant, and the system can be regarded as being in an almost steady state. As shown in Fig. 7.10, an indium wire (In wire: 1 mm diameter and 500 mm length) is then installed in the area along the direction z from $(x, y, z) = (0, 0, 0)$ to $(x, y, z) = (0, 0, 500)$ at the location of the center of the x–y plane of the system. The reaction rate distribution corresponding to the thermal neutron flux is measured in the direction z at the center of the system. Note that the In wire was chosen as the sample for the measurement because the $^{115}\text{In}(n, \gamma)^{116m}\text{In}$ reactions are very sensitive to thermal neutrons and, as a result, are likely to interact with thermal neutrons. Table 7.3 shows the characteristics of In as an activation foil for thermal neutron detection.

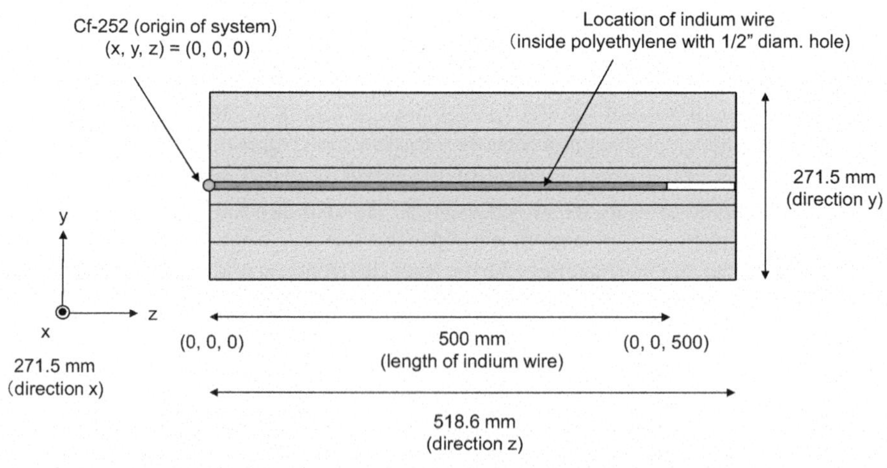

Fig. 7.10 Location of indium wire for measuring reaction rate distribution

Table 7.3 Characteristics of activation foil (In) for thermal neutron detection (Refs. [6, 7])

Nuclide	Reaction	Half-life $T_{1/2}$	Emission rate (%)	Energy of γ-ray (keV)
^{115}In	^{115}In(n, γ)^{116m}In	54.29 m	32.4	416.9
			55.7	1097.3
			85.0	1293.5

As shown in Table 7.3, when the In wire is irradiated with thermal neutrons, the saturated activity (see Chap. 5) can be determined using the information about the γ-ray energy obtained from the ^{115}In (n, γ)^{116m}In reactions. With the time correction regarding the irradiation time, the saturated activity can be treated approximately as equivalent to the reaction rate.

Without going into the derivation of the saturated activity, the saturated activity D_∞ (s^{-1}) can be obtained as follows:

$$D_\infty = \frac{\lambda_{\mathrm{In}} T_c C}{\varepsilon_{\mathrm{D}} \varepsilon_{\mathrm{E}} \left(1 - e^{-\lambda_{\mathrm{In}} T_i}\right) e^{-\lambda_{\mathrm{In}} T_w} \left(1 - e^{-\lambda_{\mathrm{In}} T_c}\right)}, \tag{7.40}$$

where λ_{In} is the decay constant of In (=ln $2/T_{1/2}$) (s^{-1}), T_c the measurement time of γ-ray by the detector (s), C the average count rate of γ-ray (s^{-1}), ε_{D} the detection efficiency of γ-ray detector, ε_{E} the emission rate of γ-ray, T_i the irradiation time (s), T_w the waiting time from the end of irradiation to the start of measurement (s).

Since it is sufficient to obtain the relative distribution of reaction rates in the direction z by the In wire from the saturated activity rather than the absolute distribution of reaction rates, the decay constant λ_{In}, detection efficiency ε_{D}, emission rate ε_{E} and irradiation time T_i can be regarded as common for all samples. Therefore, the saturated activity can be approximately expressed as follows:

$$D_\infty \approx \frac{T_c C}{e^{-\lambda_{\mathrm{In}} T_w} \left(1 - e^{-\lambda_{\mathrm{In}} T_c}\right)}. \tag{7.41}$$

Furthermore, when the weight for each sample is M, applying the weight correction using M to Eq. (7.41), the following equation can be regarded as the value corresponding to the reaction rate R:

$$R \approx \frac{D_\infty}{M} \approx \frac{T_c C}{e^{-\lambda_{\mathrm{In}} T_w} \left(1 - e^{-\lambda_{\mathrm{In}} T_c}\right) M}. \tag{7.42}$$

From the above, the relative reaction rate distribution by the In wire can be finally obtained from Eq. (7.42).

7.6 Measurements

7.6.1 Inverse Multiplication of Neutrons

Experimental steps for measuring the inverse multiplication by increasing the number of fuel assemblies are described as follows:

(1) Confirm that the number of fuel assemblies in the system is 10 ($i = 0$; Fig. 7.9a).
(2) Place the ^{252}Cf neutron source at the origin of the coordinate system,$(x, y, z) = (0, 0, 0)$ (Fig. 7.10).
(3) Confirm that ^{3}He detectors are installed at (2, G; ^{3}He#1) and (5, J; ^{3}He#2) in Fig. 7.9 ($z = 0$) and that neutrons in the system can be measured.
(4) The neutron counts at $z = 0$ mm are measured with the measurement time of 50 s, which is repeated three times, and the average of three measurements is the count rate at $z = 0$ mm. Next, move the ^{3}He#1 and ^{3}He#2 detectors from $z = 0$ to $z = 50$ mm with respect to the direction z.
(5) The neutron counts at $z = 50$ mm are measured with the measurement time of 50 s, repeated three times, and the average of three measurements is taken as the average count rate at $z = 50$ mm. Hereafter, the ^{3}He#1 and ^{3}He#2 detectors are moved in steps of 100, 150, 250, 350, and 450 mm in the direction z, and average count rates from three measurements are measured at all positions. In addition to considering the distance from the neutron source, observe well statistical errors of count rates from $z = 0$ to $z = 450$ mm measurements.
(6) The integral of the average count rate from $z = 0$ to $z = 450$ mm obtained by steps (4) and (5) is equivalent to the value of C_0 in Eq. (7.40), and this integral is approximated as C_0.
(7) Increase the number of fuel assemblies from 10 to 14 ($i = 1$; Fig. 7.9b).
(8) Repeat steps (4) and (5) to measure the value of C_1.
(9) Increase the number of fuel assemblies to 19 ($i = 2$; Fig. 7.9c) and 24 ($i = 3$; Fig. 7.9d) and measure C_2 and C_3, the values for $i = 2$ and $i = 3$, as shown in steps (7) and (8), respectively.
(10) Taking the number of fuel assemblies (10, 14, 19, and 24) on the horizontal axis of the graph and plotting C_0/C_i ($i = 0, 1, 2$, and 3) (inverse multiplication at state i: $1/M_i$) on the vertical axis, inverse multiplication curves for ^{3}He#1 and ^{3}He#2 detectors could be obtained. When obtaining the inverse multiplication curves, the data sheet in Table 7.4 can be used as a reference.

7.6.2 Reaction Rate Distribution

(1) Confirm that the number of fuel assemblies in the system is 24 ($i = 3$; Fig. 7.9d).
(2) The ^{252}Cf neutron source is set at the origin of the system,$(x, y, z) = (0, 0, 0)$ (Fig. 7.10).

Table 7.4 Datasheet for measurement of inverse multiplication (example of 24 fuel assemblies: $i = 3$)

Detector	# of fuel assemblies	Position (mm)	Counts (counts) (3 times)	Average of counts (counts)	Measurement time (s)	Average count rate (s^{-1})
He-3 #1	24	0				
		50				
		100				
		150				
		250				
		350				
		450				
				Sum of average count rates C_3 (s^{-1})		

(3) As shown in Fig. 7.10, the In wire (1 mm diameter and 500 mm length) is set in the area from $(x, y, z) = (0, 0, 0)$ to $(x, y, z) = (0, 0, 500)$.
(4) In this condition, irradiate the In wire with ^{252}Cf neutron source for about 1.5 to 2 h.
(5) After the irradiation is completed, the In wire is immediately removed from the system and cut at intervals of 50, 50, 50, 50, 70, 70, 80, and 80 mm, beginning near the ^{252}Cf neutron source, to prepare eight measurement samples.
(6) Perform γ-ray measurements on the In samples after irradiation using a high-purity germanium (HPGe) detector for all samples and record the count rate C for three energies in Table 7.3. In doing so, it is necessary to be sure to record three items: the time of the measurement T_c with the HPGe detector; the time when irradiation is stopped; and the time from the end of irradiation to the start of measurement T_w, as shown in Eq. (7.42).
(7) Measure the weight of each sample M with the use of an electronic balance.
(8) Substituting measured results obtained from steps (6) and (7) into Eq. (7.42), the saturated activity corresponding to the reaction rate can be obtained for each sample.
(9) Plotting the location of each sample from the ^{252}Cf neutron source on the horizontal axis of the graph (in order of proximity to the neutron source) and the saturated activity in step (8) on the vertical axis, the relative In reaction rate distribution corresponding to thermal neutron flux distribution in the direction z. When obtaining the distribution of In reaction rates, the following data sheet (Table 7.5) can be used as a reference.

Table 7.5 Datasheet for measurement of reaction rates (for 416.9, 1097.3 and 1293.5 keV)

Position (mm)	λ_{In} (s^{-1})	γ-ray count (counts)	t_c (s)	C (s^{-1})	t_w (s)	D_∞ (s^{-1})	M (g)	$R = D_\infty/M$ (s^{-1} g^{-1})
50								
100								
150								
200								
270								
340								
420								
500								

7.6.3 Spatial Decay Constant

(1) In the measurement of inverse multiplication, the reaction rate distributions by ^{3}He#1 and ^{3}He#2 in the direction z ($z = 0$ to $z = 450$ mm) are obtained from the results obtained in steps (4) to (9) in Sect. 7.6.1 when the number of fuel assemblies is 24. From these results, the spatial decay constants $\mu_{24}^{\mathrm{He-3\#1}}$ and $\mu_{24}^{\mathrm{He-3\#2}}$ can be determined based on Eq. (7.21).

(2) From the results of the reaction rate distribution (when the number of fuel assemblies is 24) obtained with the In wire (step (9) in Sect. 7.6.2), the captured reaction rate distribution (^{115}In(n, γ)^{116m}In) with respect to the direction z ($z = 0$ to $z = 500$ mm) is obtained. From the result, the spatial decay constant μ_{24}^{In} can be determined based on Eq. (7.21).

7.7 Discussions

7.7.1 Experimental Reports

(1) From the results of steps (4) through (9) in Sect. 7.6.1, obtain the reaction rate distributions measured from ^{3}He#1 and ^{3}He#2 at the number of fuel assemblies, 10, 14, 19, and 24. Using the results, the findings are the spatial decay constants $\mu_{10}^{\mathrm{He-3\#1,\#2}}$, $\mu_{14}^{\mathrm{He-3\#1,\#2}}$, $\mu_{19}^{\mathrm{He-3\#1,\#2}}$, and $\mu_{24}^{\mathrm{He-3\#1,\#2}}$ for 10, 14, 19, and 24 fuel assemblies, respectively, by the least-squares method. In terms of this, make discussions about the relationship between the number of fuel assemblies and the spatial decay constant.

(2) Using the results obtained from step (10) in Sect. 7.6.1, the findings are the inverse multiplication curves by ^{3}He#1 and ^{3}He#2. Here, make discussions about the shape of the curves obtained.

(3) When the count rate C_i, detection efficiency ε, neutron source strength S, and multiplication factor $k_{s,i}$ of the system are obtained by steps (4) to (9) in Sect. 7.6.1, Eq. (7.43) is approximated by the neutron source multiplication method (Chap. 10) as follows:

$$C_i = \frac{\varepsilon S}{1 - k_{s,i}}. \tag{7.43}$$

The detection efficiency ε and neutron source strength S are then assumed to be constant. Also, assuming that the multiplication factor $k_{s,3}$ for the system with 24 fuel assemblies ($i = 3$) is 0.50, obtain the multiplication factors $k_{s,0}$, $k_{s,1}$ and $k_{s,2}$ for systems with 10 ($i = 0$), 14 ($i = 1$), and 19 ($i = 2$) fuel rods, using the count rate C_i.

(4) The values of $k_{s,0}$, $k_{s,1}$, and $k_{s,2}$ are obtained from two neutron detectors (^{3}He#1 and ^{3}He#2), and make discussions about the difference between the results of ^{3}He#1 and ^{3}He#2.

(5) Using the spatial decay constants obtained from the results of inverse multiplication and reaction rate distribution (steps (1) and (2) in Sect. 7.6.3), $\mu_{24}^{\mathrm{He-3\#1}}$, $\mu_{24}^{\mathrm{He-3\#2}}$, and μ_{24}^{In} for 24 fuel assemblies, make discussions about the differences between the results of spatial decay constants obtained from the ^{3}He detectors (#1 and #2) and the In wire.

7.7.2 Reactor Physics Discussion

(1) Using the spatial decay constant obtained from the In reaction rate distribution for 24 fuel assemblies (step (2) in Sect. 7.6.3) μ_{24}^{In}, obtain the material buckling B_m^2 from Eq. (7.16).

(2) For the In reaction rate distribution corresponding to the thermal neutron flux distribution ($\phi_{24}^{\mathrm{In}}(z)$) in the estimation of the spatial decay constant (step (2) in Sect. 7.6.3), obtain the diffusion distance L_{24}^{In} by the In wire using Eq. (7.27) by the least-squares method.

(3) Giving the material buckling B_m^2 and diffusion distance L_{24}^{In} obtained in discussions (1) and (2) in Sect. 7.7.2, respectively and the Fermi age τ (see Table 7.1), the infinite multiplication factor k_∞ can then be obtained. Make discussions about the criticality of the experimental system made with NU fuel and polyethylene reflectors.

(4) When the system is in a critical steady state, using the diffusion equation with one-energy group, the material buckling B_m^2 can be expressed as follows:

$$B_m^2 = \frac{\nu \sum_{\mathrm{f}} - \sum_{\mathrm{a}}}{D}, \tag{7.44}$$

where ν is the average number of neutrons produced per fission, $\sum_f$ the fission cross sections, and $\sum_a$ the absorption cross sections. Here, discuss the possibility that the system is reached at a critical state, considering the composition of NU fuel and polyethylene moderator in the fuel assembly shown in Fig. 7.3.

References

1. U.S. Atomic Energy Commission, Division of Technical Information (1942) The fist reactor. Oak Ridge, Tennessee
2. Glasstone S, Edlund MC (1952) The elements of nuclear reactor theory. D. Van Nostrand Company Inc., New York
3. Atomic Nuclear Society of Japan, Professional Committee of Critical Experiments (1964) Reactor physics experiments. Corona Publishing Co. Ltd., Tokyo, Japan (in Japanese)
4. Okumura K, Kugo T, Kaneko K et al (2007) A comprehensive neutron calculation code system. JAEA-Data/Code 2007–004, Japan Atomic Energy Agency
5. Shibata K, Iwamoto O, Nakagawa T et al (2011) JENDL-4.0: a new library for nuclear science and engineering. J Nucl Sci Technol 48:1–30. https://doi.org/10.1080/18811248.2011.9711675
6. Gilmore G, Hemingway JD (1995) Practical gamma-ray spectrometry. John Wiley & Sons Ltd., Chichester, England
7. Firestone RB, Baglin CM, Chu SYF (eds) (1998) Table of isotopes, 8th edn. Wiley, New York

Chapter 8
Measurement of Uranium Enrichment

Abstract In this chapter, a brief explanation of safeguards is given prior to the measurement of uranium (U) enrichment, together with the background of safeguards, and the identification of nuclear fuel materials, ^{235}U, ^{238}U, and thorium-232, is described. Furthermore, detailed methods for discriminating and measuring nuclear fuel materials are presented.

Keywords Uranium enrichment · Highly-enriched uranium · Depleted uranium · Safeguards · Nuclear fuel materials

8.1 Background

Nuclear facilities that handle nuclear fuel materials, including uranium (U) isotope enrichment facilities, nuclear fuel fabrication facilities, post irradiation testing facilities, and reprocessing facilities, require extremely accurate enrichment measurements. Under the circumstances of ratification of the Treaty on Non-Proliferation of Nuclear Weapons (NPT), measurement of uranium enrichment is one of the important items for the inspection analysis in safeguards.

We will review how the U fuel is produced in each step of the nuclear fuel cycle. To produce the low-enriched uranium fuel, natural uranium is mined as U ore, refined and converted to uranium hexafluoride (UF_6), and separated into enriched and depleted U fuel through an enrichment process. The enriched U fuel then undergoes a molding and fabrication process in which it is reconverted to uranium dioxide (UO_2) to manufacture fuel assemblies for light-water reactors (LWRs). When the fuel assemblies are loaded in LWRs, plutonium (Pu) by ^{238}U capture reactions is generated, and is extracted as spent fuel. The spent fuel undergoes dissolution, separation and extraction, and refining processes in a reprocessing facility to be impaired U, Pu, and radioactive waste. As described above, it is easy to imagine that strict U enrichment control is required in the nuclear fuel cycle from the production of U fuel to the final disposal of radioactive waste.

C. H. Pyeon et al., *Reactor Laboratory Experiments at Kyoto University Critical Assembly*, https://doi.org/10.1007/978-981-97-8070-9_8

8.2 Safeguards

Safeguards are validation systems to confirm that the use of nuclear materials, including U, Pu, and other nuclear materials, is limited to peaceful purposes, and not diverted to nuclear explosive devices such as nuclear weapons or other military purposes, and that there are no undeclared nuclear materials (Ref. [1]).

The NPT stipulates that non-nuclear weapons states parties to the NPT must conclude a comprehensive safeguards agreement with the International Atomic Energy Agency (IAEA) and accept IAEA's safeguards inspections. The parties to the agreement and the IAEA are to confirm that there are no errors in the accounting and control reports indicating where and how much nuclear materials are present, and to conduct inspections of facility sites as necessary to determine the state of control. For more information on safeguards, please refer to Refs. [1, 2].

The main items to be proved in safeguards are as follows:

- Equipment and facility design information.
- Equipment and facility operating information.
- Research conducted at facilities.
- Handling of nuclear fuel materials.
- Amount of nuclear fuel material.
- Types of nuclear fuel materials.

Inspections are conducted to check whether these items are as reported and to prevent non-peaceful use. The following items are specifically checked during inspections:

- As a check of the inventory of nuclear fuel materials, the ledger and the actual inventory are verified.
- The type of nuclear fuel material is confirmed by determining U, Pu, and thorium (Th), and the U enrichment is measured.
- The facility's design information is reviewed.

The presence or absence of undeclared activities is verified. This is a type of inspection known as the complementary access, which includes inspections of facilities and areas that do not handle nuclear fuel materials, collection of environmental samples, and interviews regarding the content of research.

8.3 Identification of Nuclear Fuel Materials

8.3.1 Identification of Nuclides

The γ-ray can be used to perform nondestructive inspection of nuclear fuel materials, since most nuclear fuel materials subject to safeguards emit γ-ray. Radioisotopes that emit γ-ray are characterized by their specific energy, and the isotopic composition

Table 8.1 Response of nuclear fuel material (nuclide) to γ-ray by nondestructive testing (Ref. [3])

Isotope	Energy (keV)	Quantity of γ-ray (g^{-1} s^{-1})	Mean free pass (mm)	
			High-Z	Low-Z
^{234}U	120.9	9.35×10^4	0.23	69
^{235}U	143.8	8.40×10^3	0.36	73
	185.7	4.32×10^4	0.69	80
^{238}U	766.4	2.57×10^1	10.0	139
	1001.0	7.34×10^1	13.3	159

Z: mass number

of nuclear fuel materials can be determined by measuring the energy and relative intensity of γ-ray. This is called the γ-ray spectroscopyγ. Combined with measurements of absolute intensities, quantitative information on the nuclear fuel material can be obtained from the γ-ray energy.

The nuclide can be identified by the γ-ray from the nuclear fuel material itself or from daughter nuclides of the nuclear fuel material. Here, the γ-ray response from nondestructive testing of U fuel (isotopes) is shown in Table 8.1.

From Table 8.1, enriched U fuel emits the γ-ray with the energy of 185.7 keV upon α-decay of ^{235}U. By measuring the γ-ray, the enrichment of ^{235}U can be verified. Next, the decay series diagrams for ^{235} U, ^{238}U, and ^{232}Th are shown in Figs. 8.1, 8.2, and 8.3, respectively. The generated isotopes can be identified from the decay series of ^{235}U and ^{238}U.

- ^{235}U: the γ-ray with the energy of 185.7 keV is emitted with α-decay, and protoactinium-231 (^{231} Pa) is produced with β-decay (Fig. 8.1).
- ^{238}U: the γ-ray with the energy of 185.7 keV is emitted with α-decay, and $^{234\,m}Pa$ is produced with β-decay (Fig. 8.2).
- ^{232}Th: the γ-ray is emitted with α-decay, and actinium-228 (^{228}Ac) is produced with β-decay (Fig. 8.3).

8.3.2 *Measurement of γ-Ray Spectrum*

We will see what kind of response we get from the γ-ray spectrumγ measurements of ^{235}U and ^{238}U with enriched U. The results of the γ-ray spectrum measurements of 93 wt% ^{235}U and 2 wt% ^{235}U are shown in Figs. 8.4 and 8.5, respectively. In the two figures, peaks at energies of 185.72 keV (^{235}U: red in Figs. 8.4 and 8.5) and 1001.0 keV (^{238} U: green in Figs. 8.4 and 8.5) can be seen.

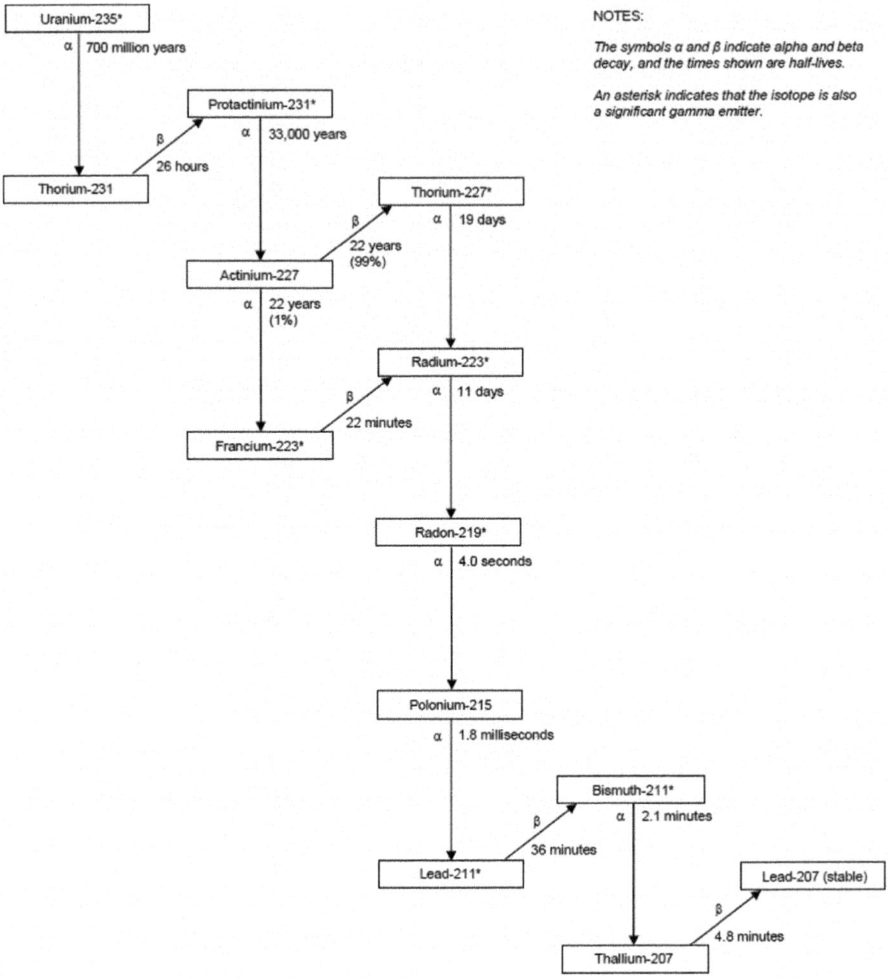

Fig. 8.1 Decay series for ^{235}U (Fig. N.2 in Ref. [4])

8.4 Experiments

8.4.1 Outline

The outline of measurements of U enrichment for nuclear fuel material is shown as follows:

(1) Set the γ-ray detector, a sodium-iodine (thallium) (NaI(Tl)) detector (or high-purity germanium: HPGe detector), to calibrate the energy of the γ-ray spectrum.

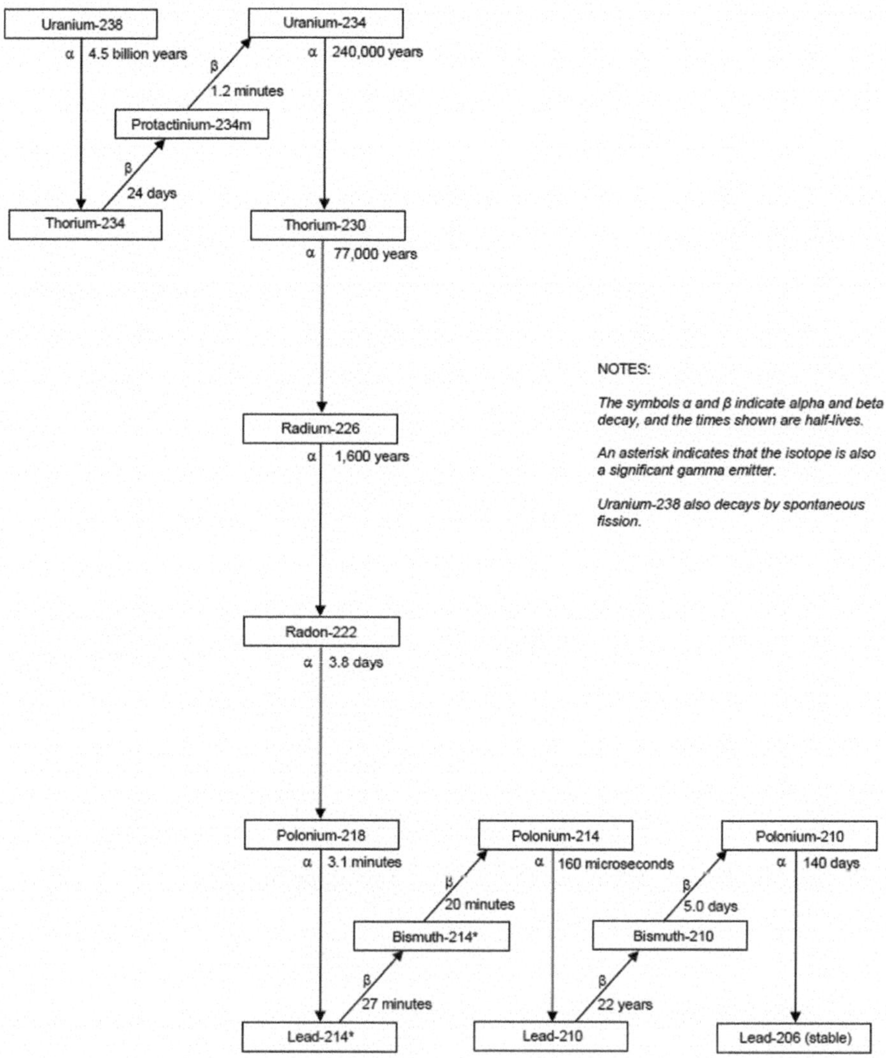

Fig. 8.2 Decay series for ^{238}U (Fig. N.1 in Ref. [4])

(2) Measure the γ-ray spectra of U samples (highly-enriched uranium: HU and depleted uranium: DU) to learn about the relationship between ^{235}U enrichment and γ-ray spectrum.

(3) Deduce the enrichment of the U sample based on the data obtained, when carrying out the γ-ray spectrum measurements with respect to U samples whose enrichment is unknown.

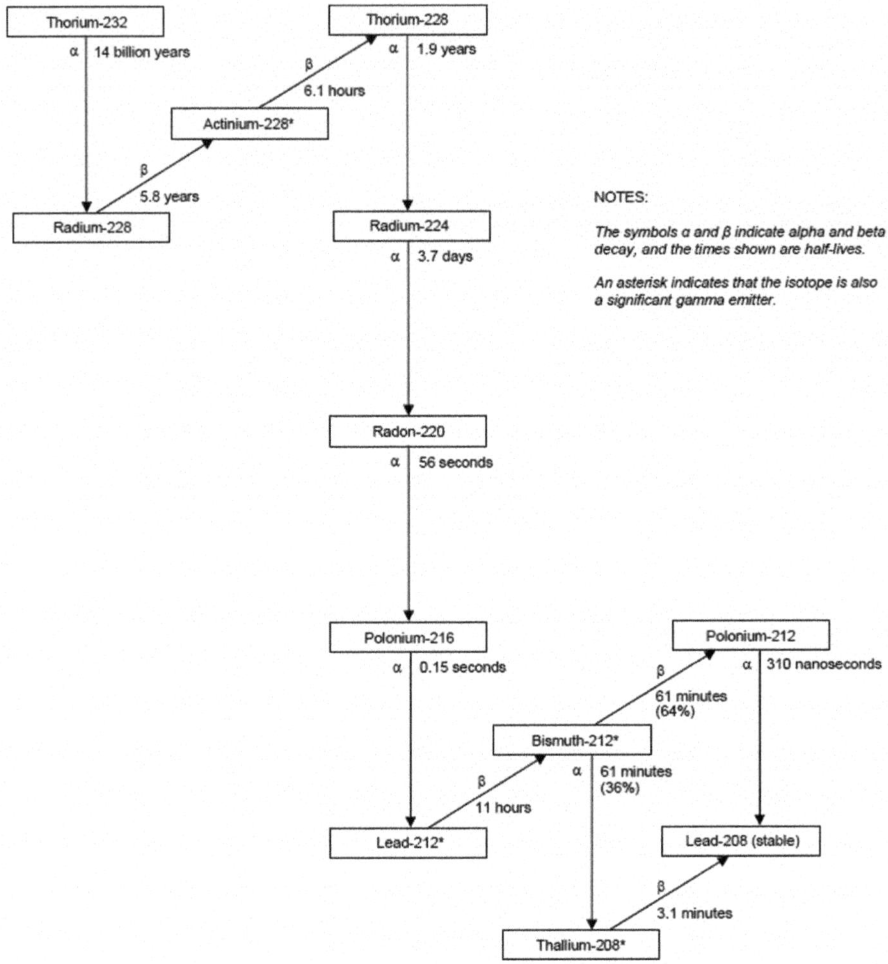

Fig. 8.3 Decay series of ^{232}Th (Fig. N.3 in Ref. [4])

(4) Perform the γ-ray spectrum measurements of ^{232}Th samples and make observations on the differences from those of ^{235}U samples (this is an optional measurement).

8.4.2 *Measurement of γ-Ray Spectrum*

An example of simultaneous measurement of two U samples, HU and DU, is shown in Fig. 8.6.

The presence of ^{235}U at 185.9 keV is identified as shown in Fig. 8.6. Also, the presence of ^{238}U is obtained from the decay of $^{234\,\mathrm{mPa}}$ at 1001.0 keV, and several

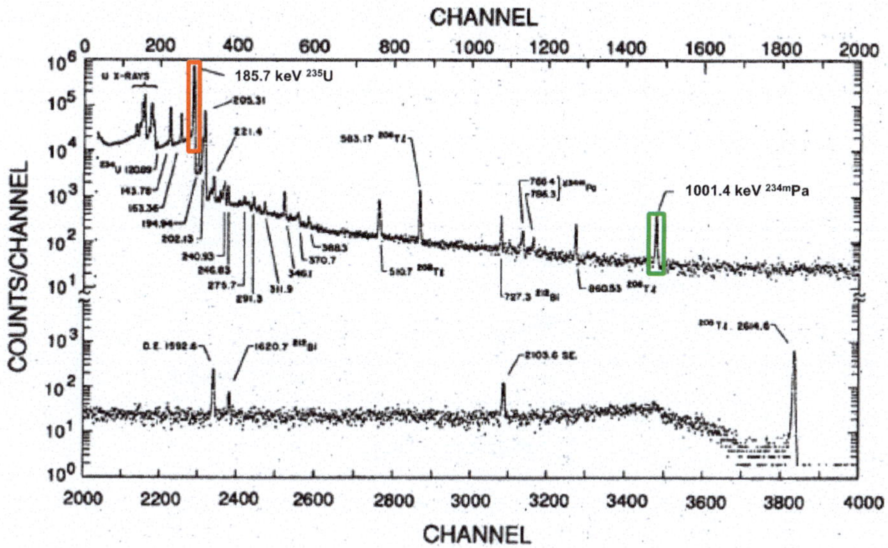

Fig. 8.4 Measured γ-ray spectrum of 93 wt% ^{235}U (Fig. 1.7 in Ref. [5])

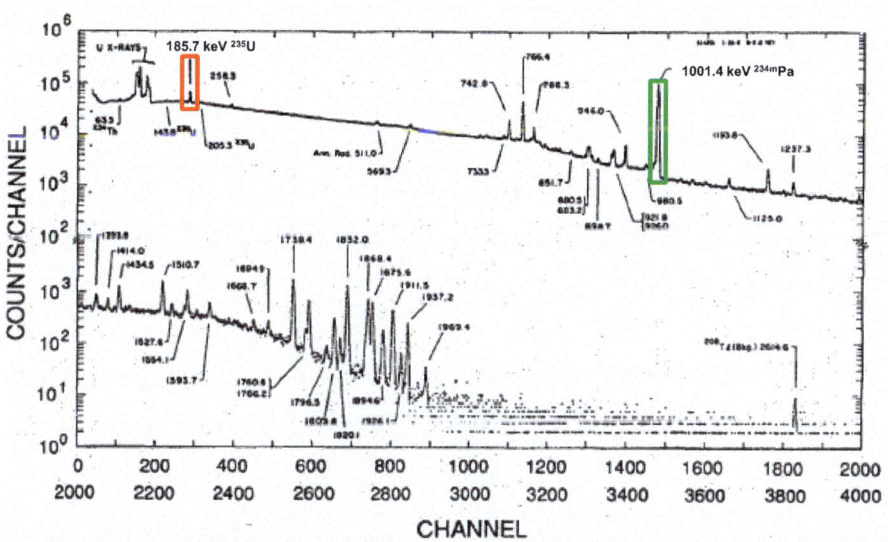

Fig. 8.5 Measured γ-ray spectrum of 2 wt% ^{235}U (Fig. 1.9 in Ref. [5])

peaks lower than 800 keV are measured as background in the γ-ray spectrum in the absence of 90 wt% ^{235}U sample (^{238}U is dominant in 0.7 wt% ^{235}U).

An example of simultaneous measurements of γ-ray spectra of two U samples (enrichment: 93 wt% and 5 wt%) is shown in Fig. 8.7. The presence of ^{235}U can

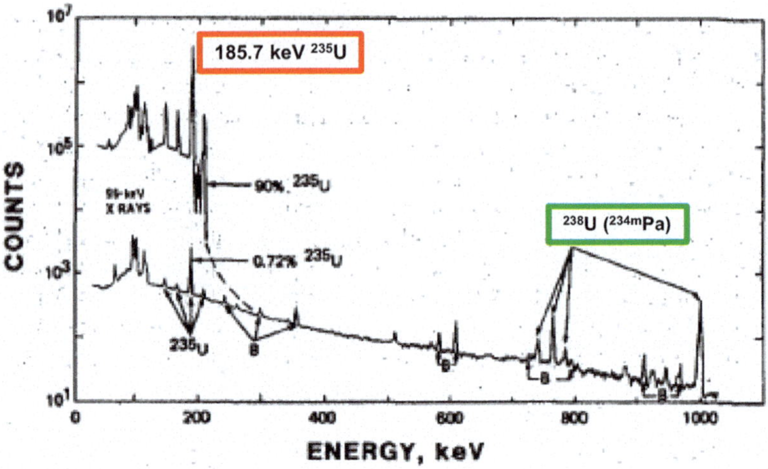

Fig. 8.6 Example of simultaneous measurement of γ-ray spectra of two U samples (90 wt% ^{235}U and 0.7 wt% ^{235}U) (Fig. 7.1 in Ref. [5])

be identified from the result of the steep peak at 185.9 keV energy (red), and the background effect of the ^{238}U daughter nuclide (green) can be observed from the result more than 200 keV.

Fig. 8.7 Example of simultaneous measurement of γ-ray spectra of two U samples (93 wt% and 5 wt% ^{235}U) (Fig. 7.2 in Ref. [5])

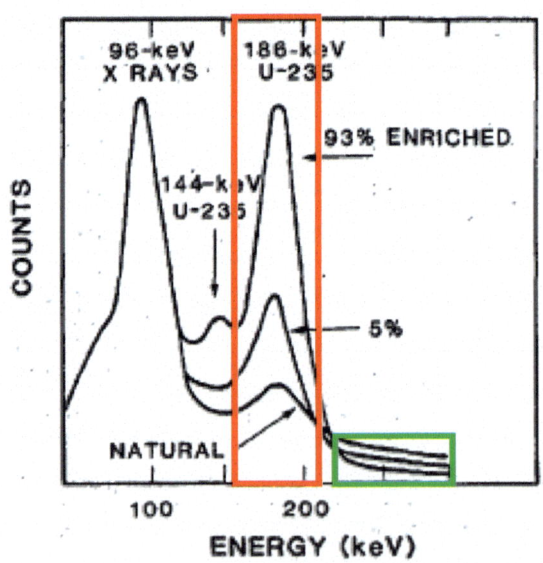

8.4.3 Simplified Method for Enrichment Measurement

This section describes the method for measuring the enrichment of U sample. If the values of the enrichment of two HU and DU samples are both unknown, the γ-ray spectrum was assumed to be obtained by simultaneous measurements of HU and DU, as shown in Fig. 8.8.

The C_1 and C_2 in Fig. 8.8 represent the 185.9 keV peak region of ^{235}U (HU) and the background (BG) region from ^{238}U (DU), respectively. Also, we assume that the enrichment of HU and DU has been known, and that two samples have the same weight. From Fig. 8.8, the BG component in the total count C_1 of HU can be determined by the BG component C_2 of DU, and the proportionality coefficient between C_1 and C_2 is then f (unknown).

In the peak region of 185.9 keV of HU, when excluding the BG component of HU calculated from that of DU, the net area (net counts) of γ-ray of HU, R [s^{-1}], can be expressed as follows:

$$R = C_1 - f \cdot C_2 \quad . \tag{8.1}$$

Next, since the enrichment (EN) [%] of a U sample is proportional to the count rate R of the γ-ray in Eq. (8.1), by introducing its coefficient e (unknown), EN can be expressed approximately as follows:

$$EN \approx e \cdot R = e(C_1 - f \cdot C_2) \quad . \tag{8.2}$$

Substituting C_1 and C_2 for HU and DU with known enrichment, respectively, into Eq. (8.2), two equations for the unknowns f and e are obtained as follows, where the subscripts of HU and DU are H and D, respectively:

$$EN_{\mathrm{H}} \approx e \cdot R_{\mathrm{H}} = e\left(C_{1,\mathrm{H}} - f \cdot C_{2,\mathrm{H}}\right), \tag{8.3}$$

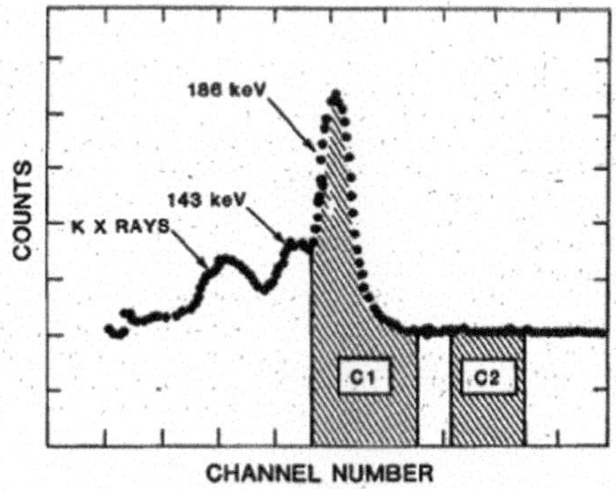

Fig. 8.8 Example of simultaneous measurement of γ-ray spectrum of two U samples (Fig. 7.4 in Ref. [5])

$$EN_{\mathrm{D}} \approx e \cdot R_{\mathrm{D}} = e\left(C_{1,\mathrm{D}} - f \cdot C_{2,\mathrm{D}}\right). \tag{8.4}$$

From Eqs. (8.3) and (8.4), f and e can be determined, respectively, as follows:

$$f = \frac{EN_{\mathrm{D}}C_{1,\mathrm{H}} - EN_{\mathrm{H}}C_{1,\mathrm{D}}}{EN_{\mathrm{D}}C_{2,\mathrm{H}} - EN_{\mathrm{H}}C_{2,\mathrm{D}}}, \tag{8.5}$$

$$e = \frac{EN_{\mathrm{D}}C_{2,\mathrm{H}} - EN_{\mathrm{H}}C_{2,\mathrm{D}}}{C_{1,\mathrm{D}}C_{2,\mathrm{H}} - C_{1,\mathrm{H}}C_{2,\mathrm{D}}}. \tag{8.6}$$

8.4.4 Data Processing

Enrichment measurements of uranium samples are performed according to the following procedures:

(1) Two γ-ray sources (cobalt-60: ^{60}Co; cesium-137: ^{137}Cs) are used to perform energy calibrations on γ-ray spectra obtained from NaI(Tl) detectors (or HPGe detectors). The γ-ray sources of ^{60}Co and ^{137}Cs have the characteristics shown in Table 8.2.
(2) The HU and DU samples with known enrichment are prepared. The two samples have the same weight. Set each sample in a NaI(Tl) detector (or HPGe detector) and measure the γ-ray spectrum, then substitute the measurement results into Eqs. (8.3) and (8.4) to determine f and e in Eqs. (8.5) and (8.6), respectively.
(3) The HU and DU samples are combined (two or three layers superimposed) to perform enrichment measurements for the two cases in Table 8.3. The datasheet in Fig. 8.9 should be utilized for the measurements.

Table 8.2 Characteristics of γ-ray sources of ^{60}Co and ^{137}Cs

γ-ray source	Half-life (year)	γ-ray energy (keV)	Emission rate (%)
^{60}Co	5.272	1173	100
		1333	100
^{137}Cs	30.17	662	85.1

Table 8.3 Cases of measurements for uranium samples HU and DU

Case	HU sample (number of samples)	DU sample (number of samples)
A	1	1
B	1	2

Detector:

HU: Enrichment = 93 wt%

Region	Index	MCA channel (range)	Counts	Measurement time [s]	Counting rate [s^{-1}]
C_1	^{235}U: 185.7 keV	~			
C_2	^{238}U: B	~			

LU: Enrichment = 0.3 wt%

Region	Index	MCA channel (range)	Counts	Measurement time [s]	Counting rate [s^{-1}]
C_1	^{235}U: 185.7 keV	~			
C_2	^{238}U: B	~			

Enrichment = $e(C_1 - f \cdot C_2)$

e = ____________ f = ____________

Case	HU : LU	C_1	C_2	Enrichment [wt%]
A	1 : 1			
B	1 : 2			

Fig. 8.9 Datasheet for measurement of uranium enrichment

- Perform the γ-ray spectrum measurement for Case A to obtain $C_{1,\text{ Case A}}$ and $C_{2,\text{ Case A}}$, and substitute them into Eq. (8.2) together with f and e obtained in step (2) to calculate the enrichment $EN_{\text{Case A}}$ for Case A.
- For Case B, follow the same procedure as above to calculate $EN_{\text{Case B}}$

8.5 Discussions

(1) Compare the results of $EN_{\text{Case A}}$ and $EN_{\text{Case B}}$ obtained by steps (2) and (3) in Sect. 8.4.4 with the theoretical enrichment (Case A: 46.6 wt%; Case B: 32.1 wt%) calculated from the combination of Cases A and B. Moreover, when observing a discrepancy between the results of experiments and references, discuss the reason of the discrepancy.

(2) Perform the γ-ray spectrum measurements on ^{232}Th samples using a NaI(Tl) detector (or HPGe detector) and compare them to the results of γ-ray spectrum measurements on U samples, and discuss any differences (this is an optional piece of homework).

References

1. Integrated Support Center for Nuclear Nonproliferation and Nuclear Security (2016) Handbook of international nuclear safeguards. Japan Atomic Energy Agency (in Japanese) http://www.mofa.go.jp/mofaj/gaiko/treaty/pdfs/A-S52-1271_1.pdf. Accessed 24 May 2024
2. IAEA Nuclear Security Series No. 6 (2007) Combating illicit tracking in nuclear and other radioactive material. International Atomic Energy Agency. https://www-pub.iaea.org/mtcd/publications/pdf/pub1309_web.pdf. Accessed 24 May 2024
3. Reilly D ed (2007) Passive nondestructive assay of nuclear materials, 2007 Addendum. Los Alamos National Laboratory. https://cdn.lanl.gov/files/passive-nondestructive-assay-of-nuclear-materials_68e9a.pdf. Accessed 24 May 2024
4. Peterson J, MacDonell M, Haroun L, Monette F (2007) Radiological and chemical fact sheet to support health risk analyses for contaminated areas. Argonne National Laboratory. https://remm.hhs.gov/ANL_ContaminantFactSheets_All_070418.pdf. Accessed 24 May 2024
5. Reilly D, Ensslin N, Smith Jr H (1991) Passive nondestructive assay of nuclear materials. NUREG/CR-5550, LA-UR-90-732, Los Alamos National Laboratory. https://www.nrc.gov/docs/ML0914/ML091470585.pdf. Accessed 24 May 2024

Chapter 9
Neutron Measurements

Abstract The target of measurement in reactor physics experiments using nuclear reactors is neutron emitted in fission reactions. The purpose of this chapter is to acquire knowledge and skills on the basic operational characteristics of neutron detectors, signal processing methods, and the fundamentals of data analysis.

Keywords Neutron detection · ^{3}He detector · BF_3 detector · Fission chamber · Optical-fiber-based detector · Signal processing

9.1 Operation Principle of Neutron Detectors

In this section, the principles of operation of ^{3}He proportional counters, BF_3 proportional counters, fission counters, and optical-fiber-based neutron scintillator detectors, which are neutron detectors used in nuclear reactor practices, are described.

9.1.1 Nuclear Reactions Used in Neutron Detectors

Neutrons are particles that have no electric charge and cannot ionize matter directly, so they are classified as so-called indirect ionizing radiation. Radiation detectors can be classified into two types: passive and active detectors. The passive detectors include activation foils [1] and thermally stimulated luminescence dosimeters (TLDs) [2], whose information is stored upon neutron irradiation and read out after irradiation. The active detectors immediately output their information during neutron irradiation. The latter is discussed in detail in this chapter.

In order to immediately generate signals during neutron irradiation, it is necessary to convert neutrons, which are indirect ionizing radiation, into charged particles, which are direct ionizing radiation, through some reactions. High-energy charged particles emitted from these reactions generate an electrical signal. High-energy photons such as γ and X-rays, are also the indirect ionizing radiation. They interact

C. H. Pyeon et al., *Reactor Laboratory Experiments at Kyoto University Critical Assembly*, https://doi.org/10.1007/978-981-97-8070-9_9

with electrons, and the produced high-energy electrons cause ionization and excitation of matter. On the other hand, neutrons interact with atomic nuclei and often produce high-energy-charged particles through nuclear reactions. The produced high-energy-charged particles cause ionization and excitation. In order to operate as an active detector, the nuclear reaction caused by neutrons must produce high-energy charged particles, which in turn generate electrical signals through ionization and excitation.

Neutrons produced by nuclear fission have relatively high energy on the order of MeV. Their high-energy neutrons are so-called fast neutrons. Fast neutrons are also possible to be used for neutron detection through nuclear reactions. Specifically, elastic scattering reaction of hydrogen (H) and fission reaction of uranium-238 (^{238}U), which is a so-called fertile nuclide, occur with fast neutrons. However, in a nuclear reactor where the chain reaction of fission is maintained via thermal neutrons, the target of detection is thermal neutrons (which have very small kinetic energy and reach to thermal equilibrium state). Thermal neutron detectors use nuclear reactions which are caused by thermal neutrons producing charged particles. Specifically, these reactions are the ^{3}He(n, p)^{3}H reaction, ^{6}Li(n, t)^{4}He reaction, ^{10}B(n, α)^{7}Li reaction, and the fission reaction. Table 9.1 summarizes the characteristics of nuclides used for thermal neutron detection. Figure 9.1 also shows the cross sections for these reactions.

The helium-3 (^{3}He) is one of the stable isotopes of He, with an isotopic abundance of only 0.00014%, and causes the following reactions:

$$ {}^{3}_{2}\mathrm{He} + {}^{1}_{0}n \rightarrow {}^{3}_{1}t + {}^{1}_{1}p + 765\ \mathrm{keV}. \tag{9.1} $$

This reaction has a cross section of 5330 barn for 25 meV thermal neutrons and is one of the most important ones in thermal neutron detection. The cross section for this reaction follows a 1/v rule up to about 10 keV. Since helium is a noble gas, it cannot form solid compounds with other elements, so its use is limited to gas detectors.

When ^{3}He undergoes the (n, p) reaction, energy of 765 keV is released, which is divided into the reaction products, tritium (^{3}H; t) and proton, p. If the incident neutron has no kinetic energy, these particles are emitted in opposite directions according to

Table 9.1 Characteristics of nuclides used for thermal neutron detection

Nuclide	Natural abundance (%)	Cross section (barn) at 25 meV		Released energy (MeV)	Reaction
^{3}He	0.00014	5330		0.765	^{3}He(n, p)t
^{6}Li	7.6	940		4.78	^{6}Li(n, t)α
^{10}B	19.9	3840	230	2.79	^{10}B(n, α)^{7}Li
			3610	2.31	^{10}B(n, α)^{7}Li*
^{235}U	0.72	587		~200	^{235}U(n, f)

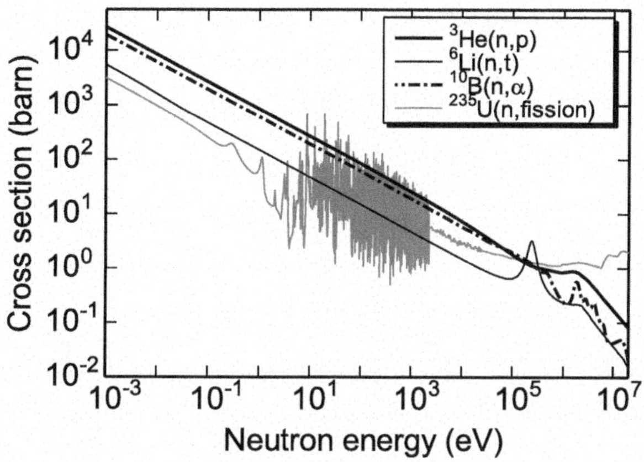

Fig. 9.1 Cross sections for reactions used in thermal neutron detectors [3]

the momentum conservation law. The kinetic energies of ^{3}He and p are $m_p/(m_t + m_p) \times 765$ keV $= 191$ keV and $m_t/(m_t + m_p) \times 765$ keV $= 574$ keV, respectively. Charged particles with these energies lose their kinetic energy as they travel through the detector medium and occur a huge number of ionizations.

The litium-6 (^{6}Li) is one of the stable isotopes of Li with isotopic abundance of 7.6%, and causes the following reaction,

$$ {}^{6}_{3}\mathrm{Li} + {}^{1}_{0}n \rightarrow {}^{4}_{2}\mathrm{He} + {}^{3}_{1}t + 4.78\,\mathrm{MeV}. \tag{9.2}$$

This reaction has a large cross section of 940 barn for a thermal neutron of 25 meV, and follows a 1/v law up to about 10 keV. The characteristics of this reaction is a much higher Q value for the reaction than that of ^{3}He. The energy is divided into the reaction products, helium ^{4}He (α-particle) and tritium t. The kinetic energy of ^{4}He and t are $m_t/(m_{4He} + m_t) \times 4.78$ MeV $= 2.05$ MeV and $m_{4He}/(m_{4He} + m_t) \times 4.78$ MeV $= 2.73$ MeV, respectively.

The boron-10 (^{10}B) is one of the stable isotopes of B with its isotopic abundance is 19.9%, and causes the following reactions,

$$\begin{aligned} {}^{10}_{5}\mathrm{B} + {}^{1}_{0}n &\rightarrow {}^{7}_{3}\mathrm{Li} + {}^{4}_{2}\alpha + 2.79\,\mathrm{MeV} \quad (6\%), \\ &\rightarrow {}^{7}_{3}\mathrm{Li}^{*} + {}^{4}_{2}\alpha + 2.31\,\mathrm{MeV} \quad (94\%). \end{aligned} \tag{9.3}$$

This reaction has a large cross section of 3840 barn for a thermal neutron of 25 meV, and follows a 1/v law up to about 100 keV. Unlike ^{3}He and ^{6}Li reactions discussed so far, this reaction produces 94% excited level ^{7}Li* and 6% ground level ^{7}Li. The excited level ^{7}Li* immediately becomes the ground level by emitting γ rays at 478 keV (= 2.79 MeV-2.31 MeV). This characteristic of γ-rays is sometimes used for neutron detection. When ^{7}Li is produced in the ground level, the Q value of the reaction is distributed $m_\alpha/(m_{7Li} + m_\alpha) \times 2.79$ MeV $= 1.01$ MeV and $m_{7Li}/(m_{7Li} + m_\alpha) \times 2.79$ MeV $= 1.78$ MeV to the reaction products ^{7}Li and α, respectively. On the other hand, if the excited level ^{7}Li* is produced, the Q value of the reaction is distributed $m_\alpha/(m_{7Li*} + m_\alpha) \times 2.31$ MeV = 0.84 MeV and $m_{7Li*}/(m_{7Li*} + m_\alpha) \times$ 2.31 MeV = 1.47 MeV to the reaction products, ^{7}Li* and α, respectively. Since ^{7}Li*, which emits γ rays, is moving at high speed, the energy of the emitted γ rays has a broadening with a center at 478 keV due to the Doppler effect. The broadening width is about 10 keV, depending on the surrounding medium. The broadening effect can be confirmed by using a high-resolution γ-ray detector such as a high-purity germanium (HPGe) detector.

The fission reaction is the most important neutron-induced nuclear reaction in the nuclear field. There are neutron detectors that use this reaction. When a heavy element such as U is irradiated by neutrons, a fission reaction might occur. This kind of reaction releases a huge amount of energy, about 200 MeV, which is distributed among the two fission fragments and two or three neutrons that are the products of the reaction. Especially, fissile nuclides, such as ^{235}U and plutonium-239 (^{239}Pu), can absorb neutrons to produce a fission reaction even if the incident neutrons are thermal neutrons, because their excess binding energy is greater than the critical energy required for a compound nucleus to fission. The fission reaction can occur even if the incident neutrons are thermal neutrons. The probability of a fission reaction for a fissile nuclide depends on the probability of forming a compound nucleus. Therefore, the reaction cross section basically follows the 1/v rule. However, it has many resonance energies and the shape indicates a complicated form.

9.1.2 ^{3}He Proportional Counters

9.1.2.1 Basic Detector Response

When ^{3}He undergoes the (n, p) reaction, energy of 765 keV is released, which is distributed to the reaction products, ^{3}H (tritium t) and proton p. If the incident neutron has no kinetic energy, these particles are emitted in opposite directions by 180 degree. In this case, the kinetic energy of ^{3}H and proton are $m_p/(m_t + m_p) \times$ 765 keV = 191 keV and $m_t/(m_t + m_p) \times 765$ keV = 574 keV, respectively. Charged particles with these energies lose their kinetic energy as they travel through the detector medium. The lost energy is used for ionization of the detector medium. As a result, a ^{3}He(n, p)^{3}H reaction produces a corresponding number of electron–ion pairs to 765 keV, which is the number obtained by dividing 765 keV by the W-value. Here, the W-value is defined as a mean production energy of a pair of electron and ion. In other words, ^{3}He(n, p)^{3}H reactions produce signals with the same pulse height corresponding to 765 keV, which is so-called as Q-value of this reaction. Consequently, the neutron peak is created in the signal pulse height spectrum, when ^{3}He proportional counter is irradiated with thermal neutrons. On the other hand, ^{3}He proportional counter has quite low sensitivity to γ rays. The probability of γ-ray interaction with the detector medium (^{3}He) is very small because ^{3}He is a gas. In the γ-ray response, the secondary electrons produced by a γ-ray interaction with the wall of the counter tube are dominant. These secondary electrons are partially reached to the detector medium gas. Because of the small stopping power of the emitted secondary electrons, the energy deposited in the γ-ray events is small compared with the neutron events. For this reason, the discrimination of the γ-ray signal from the neutron signal can be easily achieved by setting the pulse height discrimination level to an appropriate level.

9.1.2.2 Wall Effect

The small atomic number of ^{3}He results that its stopping power is small, and the range of the reaction products, *t* and *p*, may not be small compared to the detector dimensions. This causes a problem called the wall effect (Ref. [4]). The wall effect is the effect that the reaction product, *t* or *p*, reaches the detector wall material before it loses its all kinetic energy in the detector gas. In this case, a signal pulse height is lower than the pulse height corresponding to the Q value 765 keV. Figure 9.2 shows a typical pulse height spectrum obtained with a ^{3}He proportional counter affected by the wall effect.

The peak at 765 keV represents an event in which all of the energy for the Q value (765 keV) of ^{3}He(n, p) reaction is deposited into the detector gas, as described above, and is called the full-energy-absorption peak. If some of the energy of the reaction products, *t* and *p*, is absorbed by the wall, the pulse height decreases by the corresponding energy. Since energy absorbed by the wall might vary continuously, the pulse height distribution also shows a continuous distribution. However, since *t* and *p* are emitted in opposite directions by 180 degree, neither of them will be absorbed by the wall unless the detector dimensions are very small (Generally, the detector is designed to prevent such situation). In other words, in the event, in which the most energy is absorbed in the wall (i.e., the smallest pulse height), all the energy of one particle is absorbed by the wall and the other deposits all the energy into the detector gas. When *t* is absorbed by the wall (*p* is absorbe by the gas), the pulse height at the edge corresponds to 574 keV. When *p* is absorbed by the wall (*t* is

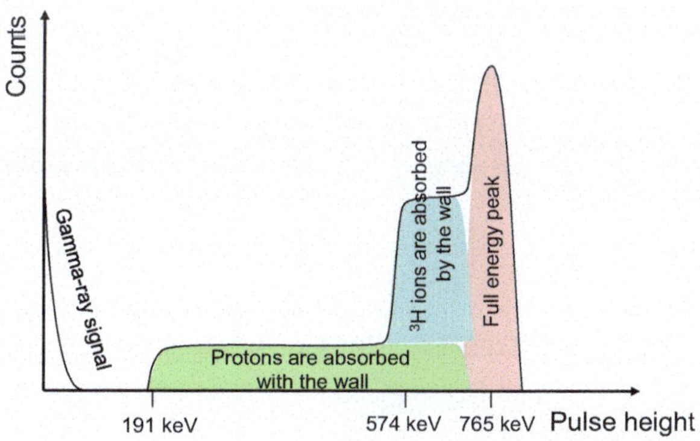

Fig. 9.2 Typical pulse height spectrum obtained with a ^{3}He proportional counter

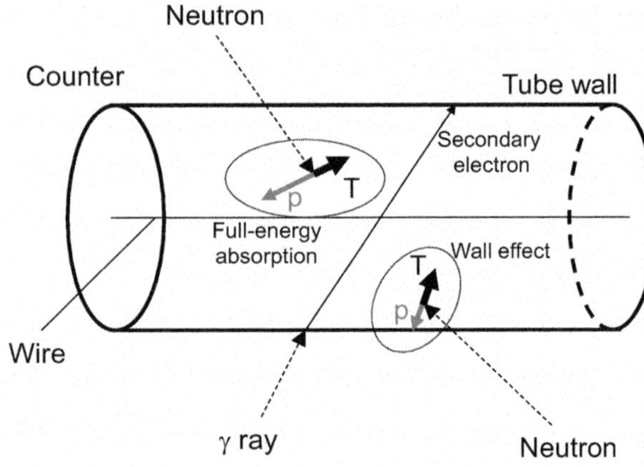

Fig. 9.3 Interactions of ^{3}He proportional counter with neutrons and γ rays

absorbed by the gas), the edge pulse height is 191 keV. As described in Fig. 9.2, a continuous distribution is added to the pulse height distribution between the full-energy-absorption peak and these edges. Such the wall effect interferes with the determination of the pulse height discrimination level, which is an undesirable effect for a neutron detector. Figure 9.3 schematically shows the interactions of the ^{3}He proportional counter with neutrons and γ rays. The range of the reaction products in ^{3}He gas at 1 atm is about 50 mm for a 574 keV proton p and about 20 mm for a 191 keV tritium t. To reduce the influence of the wall effect, it is effective to increase the detector dimensions or the gas pressure to reduce the range of the reaction products. Consequently, the ratio of the wall effect to the total absorption peak will be reduced. ^{3}He counters with a pressure from several to 10 atm or higher are commercially available. The similar results can be achieved by adding a few 10% of krypton (Kr), xenon (Xe), or carbon-tetrafluoride (CF_4) gas, which has a higher atomic number than that of He. The 574 keV protons have a range of about 1 mm when using a counting gas with a ratio of ^{3}He: $CF_4 = 30: 70$ and a pressure of 4 atm.

9.1.2.3 Detection Efficiency

The detection efficiency can be calculated as below. Since the microscopic absorption cross section of ^{3}He is 5330 b for thermal neutrons ($E_n = 25$ meV), the macroscopic absorption cross section $\sum_a$ can be expressed as follows,

$$\Sigma_a(\text{cm}^{-1}) = 5330 \times 10^{-24}(\text{cm}^2) \times N_{^3\text{He}}(\text{cm}^{-3}), \tag{9.4}$$

where $N_{3\text{He}}$ is the atomic number density of ^{3}He. When the partial pressure of ^{3}He is 1 atm ($N_{3\text{He}} = 2.7 \times 10^{19}$ cm^{-3}), the detector length L is 20 cm and the neurons are incident parallel to the axis of the counter tube, the detection efficiency ε can be calculated.

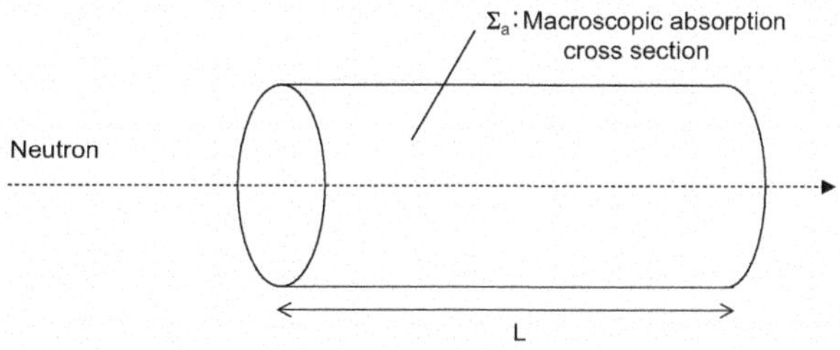

Fig. 9.4 Geometrical illustration for evaluating detection efficiency

$$\begin{aligned} \varepsilon &= 1 - \exp\left(-5330 \times 10^{-24} \times 2.7 \times 10^{19} \times 20\right) \\ &= 0.94. \end{aligned} \tag{9.5}$$

In this case, the efficiency is calculated to be 94%.

If the neutrons are incident perpendicular to the counter axis, the situation becomes more complex. The approximate efficiency can be determined by considering the effective thickness of the counter relative to the direction of neutron incidence. Monte Carlo simulations should be performed to accurately evaluate the detection efficiency (Ref. [5]) (Fig. 9.4).

9.1.3 *BF_3 Proportional Counters*

Boron can exist stably as a gas in the form of boron trifluoride (BF_3). BF_3 is often used as a detector gas in gas-filled neutron detectors. In this case, the wall effect is also a problem as in the case of ^{3}He counting tubes (Ref. [6]). Figure 9.5 shows the typical pulse height spectrum obtained from the BF_3 proportional counter when the wall effect is large. However, since the gas pressure cannot increase as in the ^{3}He counter, the range of a 1.78 MeV α particle at a typical gas pressure (0.5 atm) is about 8 mm. Therefore, the wall effect cannot be ignored in a counter with a typical tube diameter. Since charge collection efficiency becomes poor at high gas pressures due to recombination of electron and ion pairs, formation of negative ions, etc. (Ref. [7]), gas pressures of about 0.2 to 1 atm are generally used. Since the detector performance is also easily degraded by the impurity gas release from the wall material and by the dissociation of BF_3 gas, we should take care when using the BF_3 counter. In some cases, the counters can even become unusable.

The most important advantage of the BF_3 counter is its high γ-ray discrimination ability due to the larger Q-value of the ^{10}B(n, α) reaction compared to the ^{3}He(n, p) reaction. Even when the wall effect is significant, its lowest deposited energy is 840 keV, which is higher than the Q value of 765 keV for the ^{3}He (n, p) reaction. Furthermore, since the gas pressure of BF_3 is generally lower than that of ^{3}He counter, the secondary electrons generated by the interaction of γ-rays and the wall of the counter tube show low stopping power in the detector gas. The pulse height of the

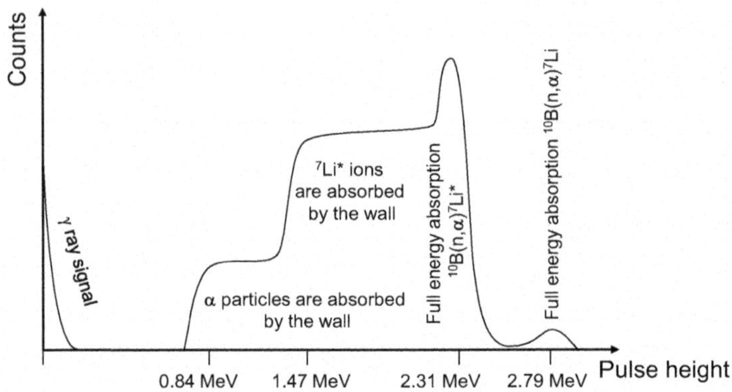

Fig. 9.5 Typical pulse height spectrum obtained from BF_3 proportional counter when the wall effect is large

γ-ray events is relatively small. Consequently, a higher γ-ray discrimination ability can be achieved compared with ^{3}He counter.

The detection efficiency of the BF_3 counter can be estimated with the same manner described in the ^{3}He counter section. For thermal neutrons (25 meV), the detection efficiency of a 20 cm detector filled with 0.5 atm of BF_3 gas using 90% enriched ^{10}B is calculated as follow:

$$\begin{aligned} \varepsilon &= 1 - \exp\left(-3840 \times 10^{-24}(\mathrm{cm}^2) \times 1.2 \times 10^{19}(\mathrm{cm}^{-3}) \times 20\,(\mathrm{cm})\right) \\ &= 0.60. \end{aligned} \quad (9.6)$$

The detection efficiency is calculated to be 60%.

9.1.4 *Fission Counters*

Fission counters usually consist of a thin layer of fissile nuclides (e.g., ^{235}U) coated on the surface of inner wall of the counter. The detection efficiency depends on the thickness of the thin layer. Generally, a thick layer detector shows high detection efficiency. However, the charge of a fission fragment is quite high (+15 k– + 20 valence), and its stopping power is huge and its range is quite short. Therefore, when the thickness of the ^{235}U layer is greater than the range of the fission fragments, the detection efficiency will not increase. In addition, since the ^{235}U layer will become a neutron absorber and make the detection efficiency low, the practical limit of the layer thickness ranges about 2 and 3 mg cm^{-2}. At this layer thickness, the detection efficiency of a single-layer structure is about 0.5% for thermal neutrons. To increase detection efficiency, some fission counters have a multilayer structure.

The most important advantage of fission counters is the high Q-value of the fission reaction, which is about 200 MeV, compared to other neutron induced reactions. Fissile materials and their daughter nuclides emit various types of ionizing radiation such as α-, β-, and γ-rays. However, they can be discriminated from the neutron events by simple pulse height discrimination due to the high Q value of the fission reaction. Of course, this high Q-value is not only applicable to γ-rays emitted from inside the detector, but also to γ-rays coming from the surroundings. Consequently the fission counter can ignore the effect of γ-rays unless the γ exposure is extremely high.

9.1.4.1 Pulse Mode

Here, we discuss the signal pulse height spectrum obtained from the fission counter in detail. Generally, the stopping power of a charged particle increases with decreasing its energy. Fission fragments have a large charge and a very high stopping power immediately after emission. However, when they are slowed down, they pick up surrounding electrons, their charge gradually decreases and their stopping power also decreases. The energy loss in the fissile material thin layer is relatively large. When the stopping power is extreme high, the correlation between the energy deposition in the detector medium and the output signal pulse height might be nonlinear. The signal pulse height is smaller than that predicted from the energy of fission fragments. However, since the energy deposition in the neutron events is quite huge compared to that in the γ-ray events, its γ-ray discrimination performance is higher than that of other neutron detectors, even taking the signal pulse nonlinearity into consideration.

9.1.4.2 Campbelling Mode (Mean Square Voltage Mode)

In very high dose fields (neutrons and γ rays), such as in nuclear reactors, individual signal pulses cannot be counted separately, and the fission ionization chamber is used in current mode, which no longer eliminates the γ-ray signal by pulse height discrimination technique. However, a mode of operation called Campbelling mode (Mean Square Voltage mode: MSV mode) can be used to reduce the influence of γ-rays (Ref. [8]). In this operation mode, the signal output is proportional not only to the number of events per unit time but also to the square of the charge produced by each event, so that the contribution of the signal due to the fission reaction, which produces a larger amount of charge (ionization number) per event, is larger than that of the γ-ray signal. Therefore, the influence of γ-rays can be suppressed. Also, even in environments, where the event rate is quite high and the pulsed mode cannot be used, fission reaction-based detectors can suppress the influence of γ rays. This feature is based on the Campbelling mode. In fact, this operation mode is widely used in neutron detectors in nuclear reactors.

9.1.4.3 Operating Principle of Campbelling Mode

If the event rate is r and the charge produced by each event is Q, then the current I, which is defined as the charge flowing through the detector per unit time, is expressed as follow,

$$I = rQ. \tag{9.7}$$

Let $n = rT$ the average number of events that occur in the detector during the effective measurement time T. Since the number of events occurring in a detector follows Poisson statistics, the standard deviation σ_n of the number of events observed during the measurement time T is written as,

$$\sigma_n = \sqrt{n} = \sqrt{rT}. \tag{9.8}$$

Therefore, the variance ${\sigma_I}^2$, which is the square of the standard deviation of the current flowing through the detector, is

$$\overline{\sigma_I^2} = \left(\frac{\sigma_n}{T}Q\right)^2 = \frac{rQ^2}{T}. \tag{9.9}$$

The variance of the current flowing through the detector, ${\sigma_I}^2$, can also be expressed from its definition as follows:

$$\overline{\sigma_I^2} = \frac{1}{T}\int\limits_{t-T}^{t} (I(t) - I_0)^2 dt, \tag{9.10}$$

where $I(t)$ is the current at time t and I_0 is the average current. In other words, by calculating the time average of the square of the AC component of the current $[I(t) - I_0]$, we can obtain an signal output that is proportional to the event rate r of a given reaction in the detector and the square of the charge Q^2 produced by that reaction. In Campbelling mode, the signal output is the variance ${\sigma_I}^2$ of the current, i.e., the time average of the square of the AC component of the current.

Let r_g and Q_g be the event rate and the charge in the fission chamber at which the signal is produced by γ-rays, respectively. Also, let r_n and Q_n be the event rate and the charge produced by neutron induced fission reaction, respectively. The signal output in current mode is

$$I = r_\gamma Q_\gamma + r_n Q_n \tag{9.11}$$

On the other hand, the signal output of the Campbelling mode is

$$\overline{\sigma_I^2} = \frac{r_\gamma Q_\gamma^2}{T} + \frac{r_n Q_n^2}{T}. \tag{9.12}$$

The ratio of contribution of γ-ray to neutron-induced signal in current mode is

$$R_I = \frac{r_\gamma Q_\gamma}{r_n Q_n}. \tag{9.13}$$

On the other hand, the ratio of γ ray to neutron in the Campbelling mode is

$$R_{\overline{\sigma_I^2}} = \frac{r_\gamma Q_\gamma^2}{r_n Q_n^2}. \tag{9.14}$$

In the fission chamber, since $Q_n > > Q_\gamma$, the γ-ray contribution can be reduced by the Campbelling mode.

9.1.5 Optical-Fiber-Based Neutron Detectors

9.1.5.1 Neutron Scintillators

A scintillator is a material that emits visible photons through the interaction of ionizing radiation and matter. Ionizing radiation quantum has the ability to occur a huge number of ionization in a matter directly or indirectly due to its high energy. In many cases, this ionization process is accompanied by excitation. Among the luminescence phenomena that occur during the relaxation process from this excited state, the scintillation process is defined as the luminescence with a relatively short decay time constant of the relaxation process. Indirect ionizing radiation such as neutrons cannot excite the scintillator directly. Therefore, it is necessary to produce high-energy charged particles, which are direct ionizing radiation, through some nuclear reactions. Thus, a material that converts neutrons (indirect ionizing radiation) into high-energy charged particles (direct ionizing radiation) is necessary to be implemented in neutron detectors. This kind of material is called a neutron converter. The neutron converter in the neutron scintillator requires the following points.

There are several possible patterns of neutron converters. Some converter materials emit high-energy ions by reaction with neutrons, emit internal conversion electrons, emit prompt gamma rays, which produce fast electrons, and produce radioactive isotopes (so-called activation), which might emit β rays. When detectors require fast response, the converter producing radioactive isotopes is not suitable because we should wait radioactive disintegration. In addition, the γ-ray-emitting converter is not suitable for effective detection due to a high penetrating power of γ rays. In this case, the large scintillator should be used to effectively detect emitted gamma rays. When comparing converters emitting energetic ions and fast electrons, the former has an advantage. Generally, energetic ions have high linear energy transfer (LET), which allows highly-localized energy deposition. This can make a scintillator dimension (thickness) small (thin). This feature can also suppress γ-ray-induced signals. As described above, lithium and boron, which undergo the ^{6}Li(n, t)α and ^{10}B(n, α)^{7}Li reactions, are candidates as converter materials for neutron scintillators.

The energetic ions generated in these reactions have relatively high LET. High LET particles induce highly-dense excitation of the scintillator. However, this feature also causes the physical quenching phenomenon, in which excited electrons do not contribute to scintillation emission and are quenched due to the high-density energy deposition. The energetic ions produced in the ^{6}Li (n, t)α reaction have a lower LET than those produced in the ^{10}B(n, α)^{7}Li reaction because they are light ions and have higher kinetic energy. For this reason, many neutron scintillators use Li as a neutron converter (Ref. [9]).

9.1.5.2 Neutron Detectors Using White Scintillator and Wavelength-Shifting Fiber

The optical-fiber-based neutron scintillator detector used at Kyoto University Critical Assembly was made of a LiF/ZnS:Ag mixed powder scintillator (Ref. [10]). The ZnS:Ag is a powder scintillator with high scintillation light yield, and the LiF powder is a neutron converter. Both are mixed and solidified with a binder. Figure 9.6 shows the structure of the detector.

A LiF/ZnS:Ag mixed powder scintillator is coated to the side of the wavelength-shifting fiber. A wavelength-shifting fiber is an element that can collect photons from the side surface of an optical fiber. Incident photons are absorbed in the optical fiber core, and then photons wavelength-shifted to the longer wavelength side are re-emitted. A part of the re-emitted photons can be transmitted in the optical fiber in the transmission mode. Generally, the incident photons from the side surface cannot be transmitted in the optical fiber. However, by using the wavelength-shifting fiber, the photons can be collected and transmitted in the optical fiber. LiF/ZnS:Ag mixed powder scintillators have almost no transparency. If the scintillator thickness increases, the scintillation photons cannot escape out from the scintillator. Consequently, this type of scintillator has optimum thickness for effective neutron detection. The LiF/ZnS:Ag mixed powder is thinly coated on the side surface of the wavelength-shifting fiber. Since the attenuation length of the wavelength-shifting fiber is not very long, it must be connected to a normal optical fiber to transmit the

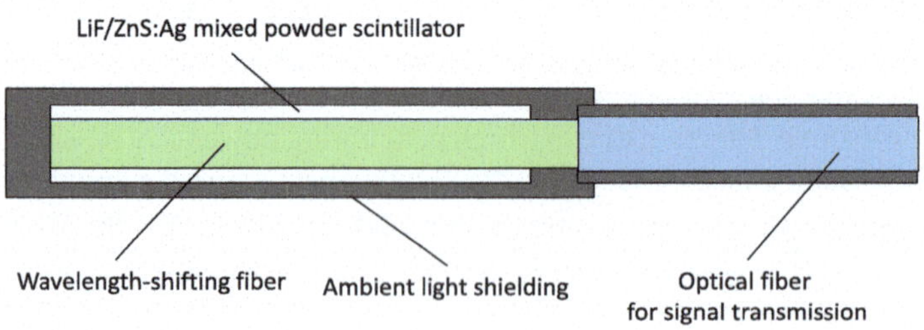

Fig. 9.6 Detector structure using LiF/ZnS:Ag scintillator and wavelength-shifting fiber

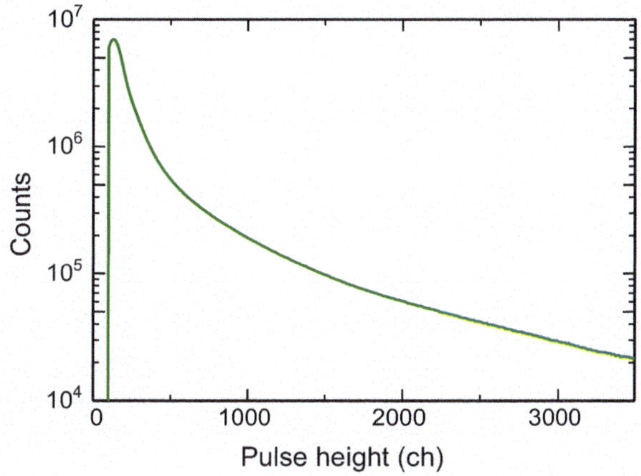

Fig. 9.7 Pulse height spectrum obtained from the neutron detector using LiF/ZnS:Ag scintillator and wavelength-shifting fiber

optical signal over a long distance. This type of neutron detectors has relatively low γ-ray sensitivity. Since the LETs of fast electrons generated by γ rays are small, the scintillation (proportional to the signal pulse height) produced in γ-ray events is kept very small compared to that produced by neutrons. As a result, it is possible to distinguish neutron events from γ-ray events by the simple pulse height discrimination. However, since the LiF/ZnS:Ag mixed powder scintillator is white and opaque, the number of scintillation photons captured by the optical fiber depends on the interaction depth and is not uniform. Consequently, the signal pulse height is not constant unlike a ^{3}He detector, but has an exponential distribution to the pulse height as shown in Fig. 9.7. As a result, there is no indicator to determine the signal discrimination level for neutron detection. It is important to set the level to prevent interference of γ-ray events.

9.1.5.3 Neutron Detectors Using Transparent Scintillator and Wavelength-Shifting Fiber

The only drawback of optical-fiber-based neutron detectors using LiF/ZnS:Ag mixed powder scintillators is that the signal pulse height spectrum has an exponential distribution. This means that it is difficult to recognize deterioration of the optical fiber, scintillator and so on, because the exponential function does not change its shape if the sensitivity and/or signal gain change. In order to recognize changes in the transmittance of an optical fiber or signal gain, it is necessary to form a characteristic shape such as a peak in the signal pulse height spectrum. One way to solve this problem is to use transparent neutron scintillators. By using transparent scintillators, scintillation light can be collected uniformly. In the ^{6}Li (n, t)α reactions, and the same number of scintillation photons are emitted from the scintillator. If scintillation photons are collected uniformly, the signals has the same pulse heights. Consequently, a peak structure is created in the pulse height spectrum. One of the

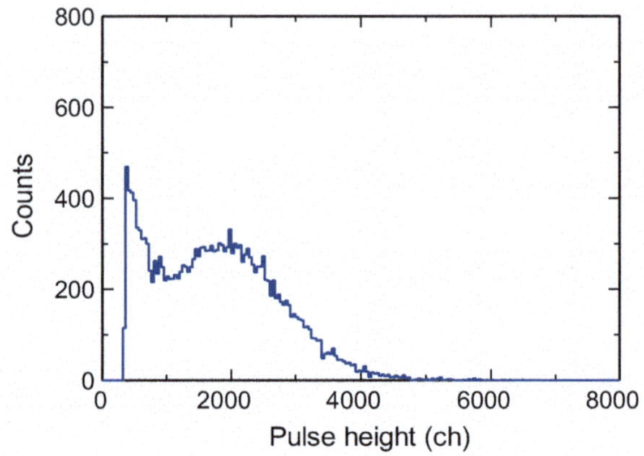

Fig. 9.8 Pulse height spectrum obtained from neutron detector using LiF/CaF_2:Eu scintillator and wavelength-shifting fiber

nearly transparent scintillator containing Li is the LiF/CaF_2:Eu eutectic scintillator (Ref. [11]). This is a eutectic crystal of LiF and CaF_2:Eu. The LiF works as a neutron converter, and CaF_2:Eu is a scintillator showing high light yield. This eutectic is formed with a mixing ratio of LiF: $CaF_2 = 80{:}20$ in mole ratio. The emission wavelength of this scintillator is 428 nm. At this wavelength, LiF/CaF_2:Eu eutectic is nearly transparent due to close refractive indices of LiF and CaF_2. The scintillation light yield is about 10,000 photons/neutron. The scintillation decay time constant is 700 ns. The density is 2.85 g/cm^3 and it has no hygroscopicity. By attaching small pieces of LiF/CaF_2:Eu to the side surface of a wavelength-shifting fiber, an optical-fiber-based neutron detector can be constructed in the same way as that using a LiF/ZnS:Ag mixed powder scintillator. Figure 9.8 shows an example of signal pulse height spectrum obtained from the optical-fiber-based neutron detector using LiF/CaF_2:Eu scintillator. The peak structure corresponding to neutron events can be confirmed. Owing to this feature, it is easy to determine the signal discrimination level to recognize neutron signals. In addition, by monitoring the peak positions, diagnosis of optical fiber degradation can be conducted.

9.2 Fundamentals of Signal Processing System

In radiation detectors, interaction events between radiation and matter result in the formation of charge signals through some processes. Usually, the generated charge signal is very small, so various signal processes are applied and the signals are finally recorded as data. Figure 9.9 shows an example of a typical signal processing circuit system. Generally, the detector and preamplifier are installed near the measurement point, and the rest of the equipment is placed in the measurement room where the measurement persons operate the instruments.

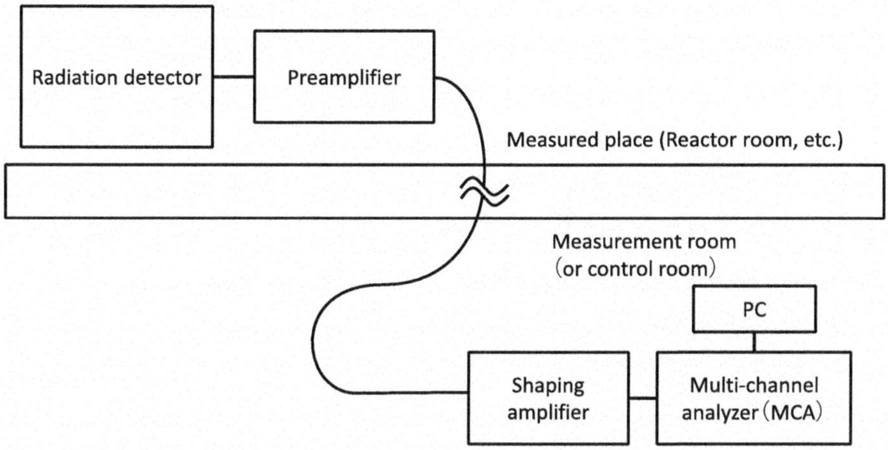

Fig. 9.9 Example of the signal processing circuit system

9.2.1 Preamplifier

The charge signals from a detector are generally quite small, and will easily be buried in various noises without any signal processing. To overcome this situation, a preamplifier is used. The main roles of a pre-amplifier are signal amplification and impedance conversion. In the preamplifier, small signals are amplified to a signal level that can be easily handled. The impedance transformation process is a bit more difficult to understand. We, therefore, will discuss it in some detail. In nuclear reactor experiments, we must take care the long-distance transmission of small signals from detectors to a control room. If long signal lines are used to connect between a detector and devices, even the resistance of the signal lines becomes a problem. A preamplifier is used to avoid this difficulty. From this viewpoint, the output impedance of the output device must be kept low. On the other hand, the input impedance of the receiving device must be kept high. The concepts of output impedance and input impedance are shown in Fig. 9.10. Impedance is the ratio of voltage to current in an electric circuit. In a DC circuit, it is equivalent to electrical resistance, and impedance is the extension concept to an AC circuit.

Considering the circuit in Fig. 9.10 as a DC circuit, Kirchhoff's law and Ohm's law give the following equations.

$$V_0 = (Z_{out} + R_{cable} + Z_{in})\, i \quad (\because \quad Z_{in} i = V_{out}), \tag{9.15}$$

$$V_{out} = \frac{Z_{in}}{Z_{out} + R_{cable} + Z_{in}} V_0. \tag{9.16}$$

When the output impedance Z_{out} of the output device becomes larger, the output voltage V_{out} becomes smaller and the difference from the source voltage V_0 becomes

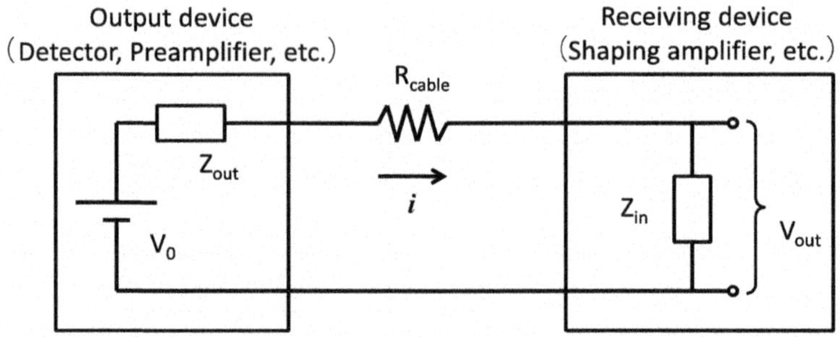

Fig. 9.10 Output impedance (Z_{out}) and input impedance (Z_{in})

larger. On the other hand, the same situation occurs when the input impedance of the receiving device becomes smaller. Therefore, it is ideal to make the output impedance of the output device small and the input impedance of the receiving device large. In such a situation, even if signal cable resistance R_{cable} is not negligible, the output voltage V_{out} will be approximately the same as the source voltage (i.e., it will be less susceptible to attenuation by the signal line).

Thus, the preamplifier has the role of impedance transformation for efficient signal transmission. Figure 9.10 is only a diagram to illustrate output impedance and input impedance. In practice, impedance transformation is performed by such an equivalent circuit. Impedance conversion is performed by field effect transistors (FETs) and similar devices in the input stage of many preamplifiers.

The charge produced in a radiation detector has information on the ionization caused in the sensitive volume of the detector, i.e., the deposition energy. By integrating the signal pulse current from the detector, the energy information deposited to the detector within each single radiation event can be obtained. Many radiation measurement systems utilize this deposited energy information. The pre-amplifier also has the function of integrating the charge from the detector. In many preamplifiers, the integrated signal is discharged with an appropriate time constant to prevent signal pile-up (overlap) and to return to the base level, after the required integration time, as shown in Fig. 9.11. The preamplifier transmits the information of the deposition energy to the detector as its signal pulse height to the signal processing device at a later stage, such as a pulse shaping amplifier.

9.2.2 Shaping Amplifier

The pulse shaping amplifiers are used to make the detector or pre-amplifier signal more manageable and to suppress signal pileup. Figure 9.12 shows the relationship between the output signals of the pre-amplifier and the shaping amplifier. The pre-amplifier integrates the detector current and output the pulse height information to the subsequent processing system. To accurately maintain and transmit the pulse

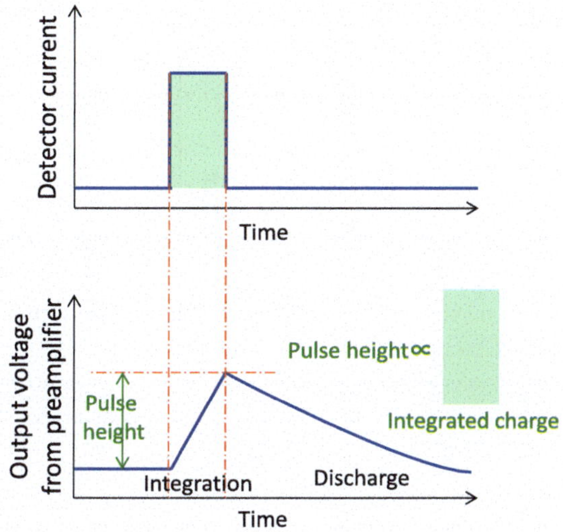

Fig. 9.11 Relationship between the detector current and the output signal from the preamplifier

height information, it has a long tail component. Therefore, when the next event occurs before this tail component returns to the base level, a signal pileup (overlap) occurs. To properly read pulse height information and to prevent signal pileup, the shaping amplifier modifies a signal shape. The shaped signal has a near flat top and the signal returns to the baseline level as quickly as possible.

Often, pulse shaping circuits, such as trapezoidal filters and Gaussian filters, are used. A typical Gaussian filter (a filter that shapes the pulse shape into a Gaussian pulse) is the CR-(RC)^n filter. It consists of a single-stage CR differential circuit

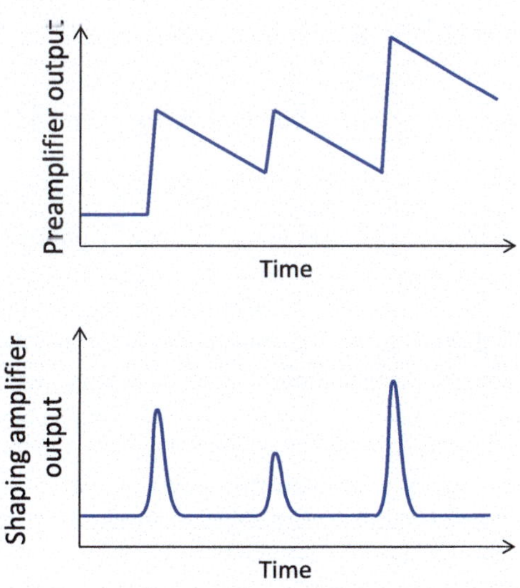

Fig. 9.12 Relationship between the output signals of the pre-amplifier and the shaping amplifier

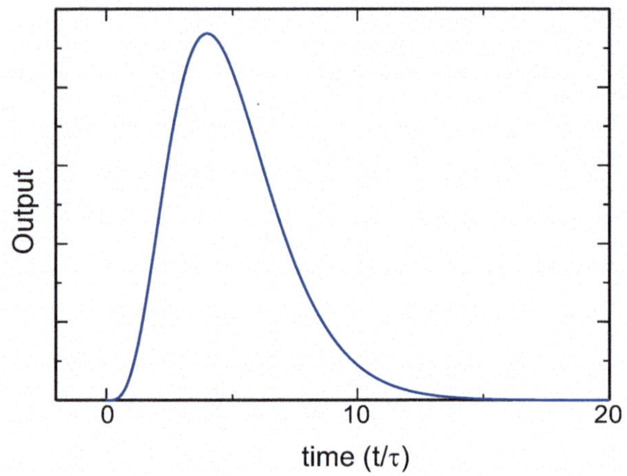

Fig. 9.13 Output pulse shape of the CR-(RC)^4 filter

followed by n-stage RC integrating circuits. Although the details are left to specialized textbooks (Ref. [12]), the output obtained when a step voltage signal is input to this filter is expressed as follows:

$$V_{out}(t) = V_0\left(\frac{t}{\tau}\right)^n e^{-\frac{t}{\tau}}, \tag{9.17}$$

where $V_{out}(t)$ is the output voltage at time t, V_0 is the signal pulse height of the input step signal, and τ is the time constant of the CR and RC circuits. Figure 9.13 shows the output pulse shape of a CR-(RC)^4 filter with four-stage integration circuits. Many shaping amplifiers use this CR-(RC)^4 filter.

9.2.3 Multi-channel Analyzer (MCA)

Multi-Channel Analyzer (MCA) is a module that processes the signal from a radiation detector (and its subsequent signal processing device, in most cases, shaping amplifier) and generates a frequency distribution (histogram) of signal pulse heights. In a conventional MCA, the pulse height values of the input analog signals are analog-to-digital converted (AD conversion), and the histogram of the digitized pulse heights is recorded in memory. In this practice, we will use a digital MCA, which has been actively used in recent years. In a conventional MCA, only the signal pulse height values are converted to digital values. On the other hand, in a digital MCA, the input analog signal waveform is first converted to digital data. From the digitized signal waveform, the pulse height information is extracted by a digital signal processing. For a digital signal processing, a special circuit called a Field-Programmable Gate Array (FPGA) is often used. In a digital signal processing, it is not difficult to calculate not only the pulse height but also the rise time and event detection time, etc. Figure 9.14

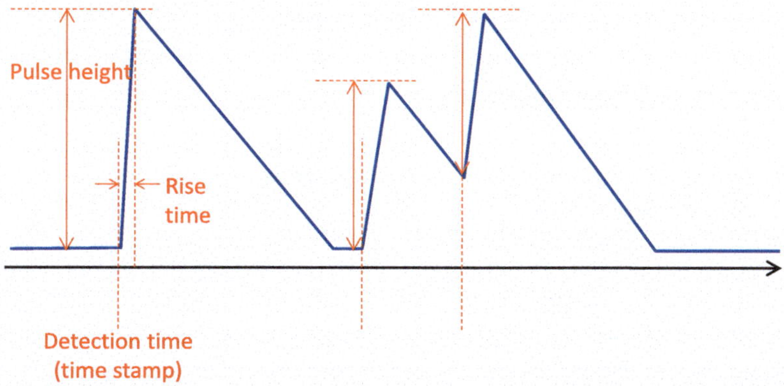

Fig. 9.14 Information extracted by a digital MCA (ANS-ZMCAN) produced by ANSeeN Inc

shows the information extracted by ANSeeN's digital MCA (ANS-ZMCAN) used in the practice.

The information extracted by the FPGA in the digital MCA is transferred to the control PC as list-mode data. List-mode data are recorded as a list of individual signal pulse information such as detection time, pulse height and rise time for each individual radiation detection event.

Figure 9.15 shows an example of list-mode data. By processing the list mode data on the control PC, a pulse height spectrum (histogram) can be created. Furthermore, if advanced processing is implemented in the processing program on the PC, it is also possible to perform coincidence measurements and create histograms of signal detection time differences for use in reactor noise analysis in reactor experiments. It is also possible to create a time trend of the signal count rate.

Event no.	Detection time	Pulse height	Input ch	Rise time
1	22573.600000	380	1	4
2	26032.700000	267	2	4
3	28471.666667	570	1	6
4	29607.766667	269	4	4
5	29977.266667	198	1	7
	⋮			

Fig. 9.15 Example of the list-mode data

9.3 Neutron Measurement Experiments

9.3.1 Experimental Method

9.3.1.1 Observation of Signals from Neutron Detectors and Signal Processing Devices

In this practice, a ^{3}He proportional counter, a BF_3 proportional counter, and a optical-fiber-based neutron scintillator detector will be used. In order to understand the function of each detector and each signal processing circuit, the signals from these neutron detectors are first observed through a preamplifier because the signal outputs of the ^{3}He and BF_3 proportional counters are very small and it is difficult to observe them directly. The output of the optical-fiber-based neutron scintillator detector, or photomultiplier tube (PMT), is relatively large and can be observed directly. These outputs will be observed with an oscilloscope. For PMT output observation, the input resistance of the oscilloscope will be set to 50 Ω or a 50 Ω terminating resistor (terminator) will be connected to the input terminal of the oscilloscope to allow observation of high-speed signal components. The signal pulse shapes will be compared between the 50 Ω and 1 MΩ input resistance conditions. In addition, the signals through the preamplifier for PMT will also be observed. Record (sketch) these signal waveforms.

Feed the output signal of the preamplifier into the shaping amplifier and observe its output signal. At this time, it is important to note the time width of the signal pulse and compare it with the preamplifier output signal. The shaping amplifier has knobs to change the "Coarse Gain," "Fine Gain," and "Shaping time." Change these knobs and record how the signal waveform changes.

9.3.1.2 Measurement of Signal Pulse Height Spectra in Various Neutron Detectors

In this section, we will measure the signal pulse height spectra of various neutron detectors to understand the principle of operation and the differences in characteristics of each detector. Feed the output of the shaping amplifier to the digital MCA to obtain the signal pulse height spectra. A californium-252 (^{252}Cf) source is used as a neutron source. The ^{252}Cf is a radioactive isotope source that undergoes spontaneous fission and emits neutrons having a fission spectrum with an average energy of approximately 1 MeV. The ^{252}Cf source is placed in a polyethylene moderator, and the generated fast neutrons are moderated to thermal neutrons. The moderated thermal neutrons are irradiated to a neutron detector, and the signal pulse height spectrum is measured using the digital MCA.

To observe not only the response to neutrons but also the response to γ rays, cesium-137 (^{137}Cs) source is used. The ^{137}Cs is a γ-ray source emitting 662 keV γ rays. The ^{137}Cs source is placed near the detector to measure the signal pulse height spectrum.

The digital MCA automatically outputs the signal pulse height spectrum in the normal measurement mode. At the same time, it also outputs list-mode data. In this practice, we will analyze the list-mode data and generate the signal pulse height spectrum in order to understand the structure of the list-mode data and how the signal pulse height spectrum is created.

9.3.1.3 Observation of Effect of Moderator

The neutron energy spectrum irradiating the detector depends on the moderator system. The thicker the moderator, the more neutrons are moderated and thermalized. As a result, the percentage of thermal neutrons increases. Since polyethylene moderators are in block form, the moderator geometry can be easily changed by rearranging them. We will reconfigure the moderator geometry in several patterns and compare the responses of various neutron detectors. If necessary, the detectors are shielded with cadmium (Cd) or lithium fluoride (LiF) plates, and the change in neutron response is observed. If available, simulations will be performed for each geometry using Monte Carlo codes such as PHITS code (Ref. [5]) to compare the results with those obtained experimentally.

9.3.2 Discussion

9.3.2.1 Observation of Signals from Neutron Detectors and Signal Processing Devices

(1) Compare the output signals from the preamplifier the shaping amplifier, and observe and measure the time width of each signal pulse. Describe the characteristics of each signal.
(2) Compare the output of a photomultiplier tube terminated in 50 Ω and that terminated in 1 MΩ, and describe the characteristics of each signal.
(3) Record how the time width of the signal pulse and the pulse height change when the time constant of the shaping amplifier is changed, and describe the characteristics.
(4) Compare the preamplifier output signals among various neutron detectors and describe the differences.

9.3.2.2 Measurement of Signal Pulse Height Spectra in Various Neutron Detectors

(1) Compare the signal pulse height spectra obtained from various neutron detectors when irradiated by thermal neutrons, and describe their characteristics, focusing on the differences between them.

(2) Compare the signal pulse height spectra obtained from various neutron detectors when irradiated by gamma rays, and describe their characteristics, focusing on the differences between them.
(3) Based on the signal pulse height spectra obtained when irradiated by thermal neutrons and gamma rays, set the signal discrimination level. The signals with pulse heights above this level is considered to be neutron signal. And state the criteria for determining the level.
(4) If possible, change the voltage applied to the detector, observe how the signal pulse height spectrum changes, and discuss the reasons.
(5) If you have more time, analyze the list-mode data and create the signal pulse height spectra.

9.3.2.3 Observation of Effect of Moderator

(1) Change the moderator geometry, record the change in the neutron signal count rate, compare how the response of each neutron detector changes, and discuss the reasons.
(2) Simulate the moderator geometry and neutron detectors using the Monte Carlo simulation code PHITS, and calculate the detector reaction rates. Compare the obtained results with the experimental results and discuss the results.

References

1. Seal MS, Mills WA, Terrill JG Jr (1957) Neutron activation analysis. Public Heal Rep 72:329–334. https://doi.org/10.4135/9781446247501.n2626
2. Oh R, Yanagisawa S, Tanaka H et al (2021) Thermal neutron measurements using thermoluminescence phosphor Cr-doped Al_2O_3 and Cd neutron converter. Sens Mater 33:2129–2135. https://doi.org/10.18494/SAM.2021.3328
3. Shibata K, Iwamoto O, Nakagawa T (2011) JENDL-4.0: a new library for nuclear science and engineering. J Nucl Sci Technol 48:1–30. https://doi.org/10.3327/jnst.48.1
4. Dietz E, Matzke M, Sosaat W et al (1993) Neutron response of an ^{3}He proportional counter. Nucl Inst Meth A 332:521–528. https://doi.org/10.1016/0168-9002(93)90309-6
5. Sato T, Iwamoto Y, Hashimoto S et al (2018) Features of particle and heavy ion transport code system (PHITS) version 3.02. J Nucl Sci Technol 55:684–690. https://doi.org/10.1080/00223131.2017.1419890
6. Nasir R, Aziz F, Mirza SM et al (2018) Experimental and theoretical study of BF_3 detector response for thermal neutrons in reflecting materials. Nucl Eng Technol 50:439–445. https://doi.org/10.1016/j.net.2017.12.014
7. Fowler IL, Tunnicliffe PR (1950) Boron trifluoride proportional counters. Rev Sci Instrum 21:734–740. https://doi.org/10.1063/1.1745700
8. Campbell NR, Francis VJ (1946) A theory of valve and circuit noise. J Inst Electr Eng 93:45–52. https://doi.org/10.1049/ji-3-2.1946.0009
9. Mori C, Osada T, Yanagida K et al (1994) Simple and quick measurement of neutron flux distribution by using an optical fiber with scintillator. J Nucl Sci Technol 31:248–249. https://doi.org/10.3327/jnst.31.248

10. Yagi T, Misawa T, Pyeon CH et al (2011) A small high sensitivity neutron detector using a wavelength shifting fiber. Appl Radiat Isot 69:176–179. https://doi.org/10.1016/j.apradiso.2010.07.016
11. Kawaguchi N, Fukuda K, Yanagida T et al (2011) Fabrication and characterization of large size $^{6}LiF/CaF_2$: Eu eutectic composites with the ordered lamellar structure. Nucl Inst Meth A 652:209–211
12. Knoll GF (2010) Radiation detection and measurement, Forth. John Wiley & Sons, Inc., New Jersey

Chapter 10
Subcriticality Measurements

Abstract The index of subcriticality is used to quantify the degree of "how deep a target system has a margin below the critical state." There are three main categories of subcriticality measurement techniques: the static method, the kinetic method, and the reactor noise analysis method. This chapter is dedicated to explaining the following methods that can be carried out at the Kyoto University Critical Assembly: the neutron source multiplication method, the inverse kinetics method, the pulsed neutron source method, and the reactor noise analysis method such as the Feynman-α method. As described in this chapter, each of the subcriticality measurement techniques has both advantages and disadvantages. Through the explanations of the principles of the measurements and the procedures, this chapter helps readers understand how to properly select the various subcriticality measurement techniques depending on an experimental condition for a target system.

Keywords Subcriticality · Neutron source multiplication method · Inverse kinetics method · Pulsed neutron source method · Reactor noise analysis method · Feynman-α method

10.1 Purpose

The purpose of this chapter is to learn expert techniques of reactor physics experiments called "subcriticality measurements" that quantify a criticality safety margin for a target system containing nuclear fuel. "Subcriticality" is an index defined by the negative reactivity $-\rho = (1 - k_{\text{eff}})/k_{\text{eff}}$, namely the degree of "how deep a target system is below the critical state ($k_{\text{eff}} = 1$)," where k_{eff} is the effective multiplication factor.

Since the dawn of reactor physics and until recently, various experimental methods of subcriticality have been proposed. Each method is classified into the following three categories:

C. H. Pyeon et al., *Reactor Laboratory Experiments at Kyoto University Critical Assembly*, https://doi.org/10.1007/978-981-97-8070-9_10

(1) Static methods to estimate the subcriticality from a measured neutron count rate at a steady state. For example, the neutron source multiplication method [1], and the exponential experiment [2].
(2) Kinetic methods to estimate the subcriticality from the time variation of a neutron count rate during the transient of the reactivity or the source strength. For example, the inverse kinetics method [3], the pulsed neutron source method [4, 5], and the source jerk method [6].
(3) Reactor noise analysis methods [7, 8] to estimate subcriticality from the statistical fluctuation of neutron counts in a steady state. For example, the Feynman-α method [9], the Rossi-α method [10], and the frequency analysis method [7, 8].

Because each of the above-mentioned methods has advantages and disadvantages, it is essential to properly use them depending on an experimental condition for a target system. Therefore, focusing on the measurement methods available at the Kyoto University Critical Assembly (KUCA), the purpose of this chapter is to promote readers' better understanding by explaining the principle and the procedure of each measurement.

10.2 Neutron Source Multiplication Method

10.2.1 Principle of Measurement

If there is no external neutron source in a target subcritical core ($k_{\mathrm{eff}} < 1$), the number of neutrons flying in the target system is also zero. In this case, when the neutron count rate measured by a neutron detector is zero, the target system can be judged to be in a subcritical state. However, the absolute value of subcriticality $-\rho$ cannot be determined from the information on the zero-neutron count rate. Thus, to investigate the subcriticality of the target system, an external neutron source should be inserted into the core. Then, fission chain reactions are initiated by neutrons emitted from the external neutron source and the source-driven subcritical system provides useful information on the non-zero neutron count rate depending on the subcriticality, the source strength, and the detection efficiency. By analyzing how many neutrons in the target core are multiplied by the subcritical fission chain reaction, the absolute value of $-\rho$ can be determined. The neutron source multiplication method is one of the static methods by which $-\rho$ can be relatively estimated from the ratio of the two neutron count rates measured in the two different subcritical systems (i.e., a reference system and the target system) at these steady states.

Let us assume that an external neutron source S is inserted into the target subcritical core. For simplicity, based on the time-dependent neutron diffusion equation with the one-energy group, the neutron flux $\phi(\mathbf{r}, t)$ can be expressed as follows:

$$\frac{1}{v}\frac{d\phi}{dt} = \nabla D(\mathbf{r},t)\nabla\phi(\mathbf{r},t) - \Sigma_a(\mathbf{r},t)\phi(\mathbf{r},t) + (1-\beta_{\text{eff}})\nu\Sigma_f(\mathbf{r},t)\phi(\mathbf{r},t) + \sum_{i=1}^{6}\lambda_i C_i(\mathbf{r},t) + S(\mathbf{r},t), \quad (10.1)$$

$$\frac{dC_i}{dt} = -\lambda_i C_i(\mathbf{r},t) + a_i\beta_{\text{eff}}\nu\Sigma_f(\mathbf{r},t)\phi(\mathbf{r},t), \quad (10.2)$$

where Σ_a and $\nu\Sigma_f$ are the macroscopic absorption and production cross sections, respectively; D is the diffusion coefficient; v is the neutron speed; β_{eff} is the effective delayed neutron fraction; C_i, λ_i, and $a_i = \beta_{\text{eff},i}/\beta_{\text{eff}}$ are the delayed neutron precursor density, the decay constant, and the relative abundance (or relative delayed neutron yield) for the i th precursor group, respectively.

After inserting the external neutron source S into the target core, it is assumed that sufficient time has elapsed to reach a steady state. Then, the delayed neutron precursor density also reaches a saturation value due to the equilibrium between the production by the fission reaction and the loss by the radioactive decay. Considering that the time derivative term in Eq. (10.2) is zero under the steady state, the spatial distribution of the stationary precursor density $C_i(\mathbf{r})$ can be obtained as follows:

$$C_i(\mathbf{r}) = \frac{a_i}{\lambda_i}\beta_{\text{eff}}\nu\Sigma_f(\mathbf{r})\phi(\mathbf{r}). \quad (10.3)$$

As can be seen from Eq. (10.3), the spatial distribution of the steady state $C_i(\mathbf{r})$ is equal to that of the fission neutron production rate $\nu\Sigma_f(\mathbf{r})\phi(\mathbf{r})$. Similarly, considering the steady state of the neutron flux (i.e., the time derivative term in Eq. (10.1) is zero and Eq. (10.3) is substituted), the steady-state diffusion equation for the source-driven subcritical system can be obtained as:

$$-\nabla D(\mathbf{r})\nabla\phi(\mathbf{r}) + \Sigma_a(\mathbf{r})\phi(\mathbf{r}) = \nu\Sigma_f(\mathbf{r})\phi(\mathbf{r}) + S(\mathbf{r}), \quad (10.4)$$

where the normalization condition of $\sum_{i=1}^{6} a_i = 1$ is used to derive Eq. (10.4).

Now let us assume that the neutron flux $\phi(\mathbf{r})$ can be expanded by the sum of the products of the eigenfunction $\Phi_n(\mathbf{r})$ in the k_{eff}-eigenvalue equation (also known as λ-eigenvalue equation) and its expansion coefficient f_n:

$$\phi(\mathbf{r}) = \sum_{n=1}^{\infty} f_n\Phi_n(\mathbf{r}), \quad (10.5)$$

where, based on the neutron diffusion theory with the one-energy group, the nth order eigenfunction $\Phi_n(\mathbf{r})$ and the eigenvalue k_n satisfy the following equations ($n = 1, 2, \cdots$):

$$-\nabla D(\mathbf{r})\nabla\Phi_n(\mathbf{r}) + \Sigma_\mathrm{a}(\mathbf{r})\Phi_n(\mathbf{r}) = \frac{1}{k_n}\nu\Sigma_\mathrm{f}(\mathbf{r})\Phi_n(\mathbf{r}), \tag{10.6}$$

$$k_n = \frac{\int_V \Phi_n^\dagger(\mathbf{r})\nu\Sigma_\mathrm{f}(\mathbf{r})\Phi_n(\mathbf{r})dV}{\int_V \left(\Phi_n^\dagger(\mathbf{r})\Sigma_\mathrm{a}(\mathbf{r})\Phi_n(\mathbf{r}) - \Phi_n^\dagger(\mathbf{r})\nabla D(\mathbf{r})\nabla\Phi_n(\mathbf{r})\right)dV}. \tag{10.7}$$

Note that the fundamental mode ($n = 1$) of the eigenvalue k_1 and the eigenfunction $\Phi_1(\mathbf{r})$ correspond to the effective multiplication factor k_eff (the ratio of the number of neutrons produced by fission to the number of neutrons lost by absorption and leakage) and the neutron flux distribution in a critical state. In the case of the one neutron energy group, the k_eff-eigenfunction satisfies the self-adjoint condition of $\Phi_n^\dagger(\mathbf{r}) = \Phi_n(\mathbf{r})$, i.e., the adjoint eigenfunction $\Phi_n^\dagger(\mathbf{r})$ is equal to the forward eigenfunction $\Phi_n(\mathbf{r})$. Furthermore, these eigenfunctions $\Phi_n(\mathbf{r})$ and $\Phi_n^\dagger(\mathbf{r})$ satisfy the following orthogonal condition using Kronecker's delta δ_{mn} to simplify the subsequent transformation of the equations:

$$\int_V \Phi_m^\dagger(\mathbf{r})\nu\Sigma_\mathrm{f}(\mathbf{r})\Phi_n(\mathbf{r})dV = \mathcal{F}_n\delta_{mn}, \tag{10.8}$$

$$\mathcal{F}_n \equiv \int_V \Phi_n^\dagger(\mathbf{r})\nu\Sigma_\mathrm{f}(\mathbf{r})\Phi_n(\mathbf{r})dV. \tag{10.9}$$

For example, the analytical solutions of the eigenvalue k_n and the eigenfunction $\Phi_n(\mathbf{r})$ for a homogeneous one-dimensional slab geometry with a thickness of H ($-H/2 \le x \le H/2$) can be solved using the square root of the geometric buckling B_n^2:

$$B_n = \frac{n\pi}{H}, \tag{10.10}$$

$$k_n = \frac{\nu\Sigma_\mathrm{f}}{\Sigma_\mathrm{a} + DB_n^2}, \tag{10.11}$$

$$\Phi_n(\mathbf{r}) = \sqrt{\frac{2}{H}}\sin\left(B_n\left(x + \frac{H}{2}\right)\right),$$

$$\text{e.g., } \Phi_1(\mathbf{r}) = \sqrt{\frac{2}{H}}\cos\left(\frac{\pi}{H}x\right)\text{for } n = 1. \tag{10.12}$$

By multiplying both sides of Eq. (10.4) by the self-adjoint eigenfunction $\Phi_n^\dagger(\mathbf{r}) = \Phi_n(\mathbf{r})$ and integrating over all space, the nth order expansion coefficient f_n can be derived from the relations of Eqs. (10.5) through (10.9):

$$f_n = \frac{k_n}{1-k_n}\frac{\mathcal{S}_n}{\mathcal{F}_n}, \tag{10.13}$$

$$\mathcal{S}_n \equiv \int_V \Phi_n^\dagger(\mathbf{r})S(\mathbf{r})dV. \tag{10.14}$$

If the subcriticality of the target system is relatively shallow to satisfy $(1-k_1)/k_1 \ll (1-k_n)/k_n$ for the higher-order modes, the steady-state neutron flux $\phi(\mathbf{r})$ can be approximated by the dominant eigenfunction of the fundamental mode $\Phi_1(\mathbf{r})$:

$$\phi(\mathbf{r}) = \frac{k_1}{1-k_1}\sum_{n=1}^{\infty}\frac{k_n}{k_1}\frac{1-k_1}{1-k_n}\frac{\mathcal{S}_n}{\mathcal{F}_n}\Phi_n(\mathbf{r}) \approx \frac{1}{-\rho}\frac{\mathcal{S}_1}{\mathcal{F}_1}\Phi_1(\mathbf{r}). \tag{10.15}$$

To derive Eq. (10.15), the equivalent condition of $-\rho = (1-k_1)/k_1$ is utilized, since k_1 corresponds to the effective multiplication factor k_{eff}.

Based on the above theoretical derivation, let us assume that an experimental condition where the fundamental mode component is dominant and a neutron detector with the macroscopic detection cross section $\Sigma_{\mathrm{d}}(\mathbf{r})$ is installed. Consequently, the measured neutron count rate R can be expressed by a function form that is proportional to the detection efficiency ε and the external neutron source strength $\mathcal{S}_1$ and inversely proportional to the subcriticality $-\rho$:

$$R = \int_V \Sigma_{\mathrm{d}}(\mathbf{r})\phi(\mathbf{r})dV \approx \frac{k_1\varepsilon\mathcal{S}_1}{1-k_1} = \frac{\varepsilon\mathcal{S}_1}{-\rho}, \tag{10.16}$$

$$\varepsilon \equiv \frac{\int_V \Sigma_{\mathrm{d}}(\mathbf{r})\Phi_1(\mathbf{r})dV}{\mathcal{F}_1} = \frac{\int_V \Sigma_{\mathrm{d}}(\mathbf{r})\Phi_1(\mathbf{r})dV}{\int_V \Phi_1^\dagger(\mathbf{r})\nu\Sigma_{\mathrm{f}}(\mathbf{r})\Phi_1(\mathbf{r})}. \tag{10.17}$$

Recalling that the target core is in a subcritical state of $k_{\mathrm{eff}} < 1$, Eq. (10.16) can also be expressed by the following sum of the geometric series with the first term $\varepsilon\mathcal{S}_1$ and the common ratio k_{eff}:

$$R \approx \varepsilon\mathcal{S}_1\left(k_{\mathrm{eff}} + k_{\mathrm{eff}}^2 + \cdots\right) = \varepsilon\mathcal{S}_1\sum_{i=1}^{\infty}k_{\mathrm{eff}}^i. \tag{10.18}$$

Based on the transformation of Eq. (10.18), the measured count rate R is proportional to the total sum of the detected descendants, which originate from the ancestor neutrons emitted by the external source and gradually decrease by the factor of k_{eff} per generation.

The subcriticality measurement technique based on the above-mentioned principle is known as the "neutron source multiplication method [1]." The detailed measurement method in the actual experiment is explained below. First, a reference subcriticality $-\rho_{\text{ref}}$ is measured in advance by some kind of method. Then, an external neutron source and a neutron detector are inserted into this reference subcritical core, and the steady-state count rate R_{ref} is measured. Thereby, based on the fundamental mode approximation shown in Eq. (10.16), the unknown εS_1 can be calibrated as follows:

$$\varepsilon S_1 = -\rho_{\text{ref}} R_{\text{ref}}. \tag{10.19}$$

Next, the subcriticality changes from the reference value of $-\rho_{\text{ref}}$ to an unknown value of $-\rho_{\text{target}}$ due to some kind of state change. Note that the positions of the external neutron source and the neutron detector remain the same and the fundamental mode of the eigenfunction $\Phi_1(\mathbf{r})$ is also unchanged after the reactivity change. Namely, the calibrated value of εS_1 is approximated to be constant before and after the subcriticality change. Then, since the measured count rate R_{target} for the steady-state target core is inversely proportional to the target subcriticality $-\rho_{\text{target}}$, R_{target} increases as the target core approaches the critical state, and vice versa:

$$R_{\text{target}} \approx \frac{\varepsilon S_1}{-\rho_{\text{target}}}. \tag{10.20}$$

Finally, based on Eqs. (10.19) and (10.20), the unknown target subcriticality $-\rho_{\text{target}}$ can be easily estimated by the steady-state count rate ratio of $R_{\text{ref}}/R_{\text{target}}$ between the reference and target states:

$$-\rho_{\text{target}} \approx -\rho_{\text{ref}} \frac{R_{\text{ref}}}{R_{\text{target}}}. \tag{10.21}$$

Note that, in the neutron source multiplication method, the product of the detection efficiency and the external neutron source strength εS_1 in the target state is approximated to be the same as in the reference state. For example, if the detection efficiency or the external neutron source is significantly changed after the state change, this approximation results in the systematic error of the estimated $-\rho_{\text{target}}$. In addition, since the neutron source multiplication method based on the fundamental mode approximation neglects the higher-order mode effect, $-\rho_{\text{target}}$ is expected to be well estimated when the subcriticality is relatively shallow. Conversely, as the subcriticality becomes deeper, the higher-order mode effect on the spatial distribution of the neutron flux increases, resulting in the larger systematic error of $-\rho_{\text{target}}$. As can be seen from the explanation of this subsection, since the neutron source multiplication method is the static method using the steady-state count rate measurements, the subcriticality cannot be accurately estimated when the count rate is measured in the time-dependent transient situation.

10.2.2 Measurement Procedure

In order to carry out the experiment of the neutron source multiplication method, the excess reactivity ρ_{excess} and the control rod worth for the ith control rod $\rho_{\text{rod},i} < 0$ are measured in advance according to the procedures as explained in Chaps. 3 and 4. Then, a reference subcritical core, whose reference subcriticality $-\rho_{\text{ref}}$ is already known, is configured by inserting and/or withdrawing the control rods. Furthermore, a target subcritical core with the unknown subcriticality $-\rho_{\text{target}}$ is also designed by changing the combination of the inserted and/or withdrawn control rods. In this case, the procedure for measuring $-\rho_{\text{target}}$ by the neutron source multiplication method is described below:

(1) Based on the previously measured excess reactivity ρ_{excess} and negative reactivity of the control rod $\rho_{\text{rod},i} < 0$, the negative reactivity of the reference core is evaluated according to the control rod pattern of the reference core. Namely, $\rho_{\text{ref}} = \rho_{\text{excess}} + \sum_i \rho_{\text{rod},i} < 0$. The absolute value of $|\rho_{\text{ref}}| = -\rho_{\text{ref}}$ corresponds to the subcriticality for the reference core.

(2) An external neutron source (e.g., americium-berylium (Am-Be) startup neutron source or californium-252 (^{252}Cf) spontaneous fission source) is inserted into the reference core with the known subcriticality. Let us allow sufficient time for the neutron counting in the source-driven subcritical core to reach the steady state.

(3) The reference value of the steady-state count rate R_{ref} is measured by a neutron detector installed in the reference core, e.g., the fission chamber (FC) for the startup channel and another inserted detector such as boron-trifluoride (BF_3) or helium-3 (^{3}He) proportional counter. For example, the measurement of neutron counts during the detection time width $T = 100$ s is repeated with the sample size of $N = 5$ to estimate the sample mean and the standard error of the mean for R_{ref}. Note that T and N should be set appropriately according to the desired measurement precision and the operation time available for the measurement.

(4) While the spatial positions of the external neutron source and the neutron detectors remain the same, the subcritical core is changed from the reference core to the target subcritical core by inserting (or withdrawing) control rods. After the reactivity transient, let us allow sufficient time for the neutron count rate to reach the steady state of the target core.

(5) In the steady-state target core, the sample mean and standard error of the mean for the neutron count rate R_{target} are estimated in the same manner as the reference core.

(6) By substituting the reference subcriticality $-\rho_{\text{ref}}$ and two measured count rates R_{ref} and R_{target} into Eq. (10.21), the unknown subcriticality $-\rho_{\text{target}}$ can be estimated.

10.2.3 Discussions

Let us discuss the measurement results of subcriticality $-\rho_{\text{target}}$ using the neutron source multiplication method, from the following viewpoints:

(1) For comparison, estimate the reference value of the subcriticality in the targe core, $-\rho_{\text{target}}$, based on the control rod pattern with the known values of the excess reactivity and the control rod worth, which were previously measured in advance.

(2) Comparing $-\rho_{\text{target}}$ estimated by the neutron source multiplication method with the reference value obtained in the previous discussion (1), investigate the measurement accuracy (or the relative difference from the reference value) to quantitatively discuss the agreement of the neutron source multiplication method. If there is a significant difference between these results, discuss the main cause of the difference based on the approximation conditions assumed in the principle of the neutron source multiplication method.

- Based on Eq. (10.15), discuss whether the subcriticality of the reference core can be considered as a near-critical system.
- Discuss the differences in the measurement results using some detectors installed at the different positions, by focusing on the positional relation among the detector, the external neutron source, and the fuel region.

(3) Discuss whether there are differences in the measured subcriticality values of $-\rho_{\text{target}}$ between the neutron source multiplication method (the static method) and other measurement methods (e.g., the kinetic method and the reactor noise analysis method as explained in Sects. 10.3 through 10.5).

(4) Based on the measured neutron count rates in the approach-to-criticality experiment as explained in Chap. 3, try to estimate the subcriticality value for each subcritical state using the neutron source multiplication method.

10.3 Inverse Kinetics Method

10.3.1 Principle of Measurement

In a critical or subcritical core, if the reactivity is changed by operating a control rod, the number of neutrons in the core will increase or decrease with respect to time, followed by the corresponding increase or decrease in the neutron count rate measured by a neutron detector. Using such information on the time variation of the neutron count rate $R(t)$ as an input value, the time variation of the reactivity $\rho(t)$ can be estimated. This experimental method is known as the "inverse kinetics method." The inverse kinetics method is one of the kinetic methods and can be utilized to monitor the reactivity change $\rho(t)$ in a target core where the strength of the external neutron source is constant.

Let us assume that a constant external neutron source S is present in a target core and the reactivity $\rho(t)$ changes with respect to time due to the control rod operation. Similar to Eq. (10.5), the time-dependent neutron flux $\phi(\mathbf{r}, t)$ can be expanded by the following sum of the products of the k_{eff}-eigenfunction $\Phi_n(\mathbf{r})$ and the corresponding time-dependent expansion coefficient $f_n(t)$:

$$\phi(\mathbf{r}, t) = \sum_{n=1}^{\infty} f_n(t)\Phi_n(\mathbf{r}). \tag{10.22}$$

By substituting Eq. (10.22) into Eqs. (10.1) and (10.2), multiplying both sides by the self-adjoint eigenfunction of the fundamental mode $\Phi_1^{\dagger}(\mathbf{r})$, and integrating over all space, the point kinetics equation for the expansion coefficient of the fundamental mode $f_1(t)$ can be derived from the relations of Eqs. (10.6) through (10.9) and (10.22). Furthermore, if the target core is close to the critical state and the fundamental mode approximation is reasonably applicable, the neutron count rate $R(t)$ can be approximated as follows:

$$R(t) = \int_V \Sigma_{\text{d}}(\mathbf{r})\phi(\mathbf{r}, t)dV \approx f_1(t)\int_V \Sigma_{\text{d}}(\mathbf{r})\Phi_1(\mathbf{r})dV = \mathcal{D}_1 f_1(t), \tag{10.23}$$

$$\mathcal{D}_1 \equiv \varepsilon\mathcal{F}_1 = \int_V \Sigma_{\text{d}}(\mathbf{r})\Phi_1(\mathbf{r})dV. \tag{10.24}$$

Consequently, the following point kinetics equation can be derived, which describes the time dependency of the measured count rate $R(t)$:

$$\frac{dR}{dt} = \frac{\rho(t) - \beta_{\text{eff}}}{\Lambda}R(t) + \sum_{i=1}^{6}\lambda_i C_i(t) + S, \tag{10.25}$$

$$\frac{dC_i}{dt} = -\lambda_i C_i(t) + \frac{a_i\beta_{\text{eff}}}{\Lambda}R(t), \tag{10.26}$$

where Λ and β_{eff} denote the neutron generation time and the effective delayed neutron fraction; and λ_i and a_i are the decay constant and the relative abundance for the ith precursor group, respectively. For example, based on the neutron diffusion theory with the one-energy group, each symbol in Eqs. (10.25) and (10.26) corresponds to the following quantity:

$$C_i(t) = \mathcal{D}_1\frac{\int_V \Phi_1^{\dagger}(\mathbf{r})C_i(\mathbf{r}, t)dV}{\int_V \Phi_1^{\dagger}(\mathbf{r})\frac{1}{v}\Phi_1(\mathbf{r})dV}, \tag{10.27}$$

$$S = \mathcal{D}_1\frac{\int_V \Phi_1^{\dagger}(\mathbf{r})S(\mathbf{r})dV}{\int_V \Phi_1^{\dagger}(\mathbf{r})\frac{1}{v}\Phi_1(\mathbf{r})dV}, \tag{10.28}$$

$$\Lambda = \frac{\int_V \Phi_1^\dagger(\mathbf{r})\frac{1}{v}\Phi_1(\mathbf{r})dV}{\int_V \Phi_1^\dagger(\mathbf{r})\nu\Sigma_f(\mathbf{r})\Phi_1(\mathbf{r})dV}. \tag{10.29}$$

If the time-dependent reactivity $\rho(t)$ and the point kinetics parameters (Λ and β_{eff}) are given as the input values for Eqs. (10.25) and (10.26), the numerical solver for the first-order differential equation such as the Euler and Runge–Kutta methods can be utilized to numerically predict the time variation of the count rate $R(t)$. By inverting this relationship, the time variation of the reactivity $\rho(t)$ can be estimated from the time dependence of increase or decrease in the count rate $R(t)$ as shown in Fig. 10.1. This estimation method is called the "inverse kinetics method" and is often utilized as the principle of measurement for real-time reactivity monitoring in actual nuclear reactors [3].

For example, to estimate the reactivity $\rho(t)$ by the inverse kinetics method, the following equation can be derived from Eqs. (10.25) and (10.26):

$$\rho(t) = \beta_{\text{eff}} + \frac{\Lambda}{R(t)}\left(\frac{dR}{dt} - \sum_{i=1}^{6}\lambda_i C_i(t) - S\right), \tag{10.30}$$

$$C_i(t) = \frac{a_i\beta_{\text{eff}}}{\Lambda}\int_{-\infty}^{t} R(t')e^{-\lambda_i(t-t')}dt'. \tag{10.31}$$

Based on Eqs. (10.30) and (10.31) with the time series data of the count rate $R(t)$ measured by a neutron detector installed in the target core, the reactivity $\rho(t)$ can be successively estimated. Compared with the neutron source multiplication method, the inverse kinetics method enables us to estimate the time-dependent $\rho(t)$ during the transient condition of the count rate $R(t)$ under not only a subcritical state but also critical and supercritical states $\rho(t) \geq 0$. Note that the target core should be in a near-critical state to reasonably apply the fundamental mode approximation.

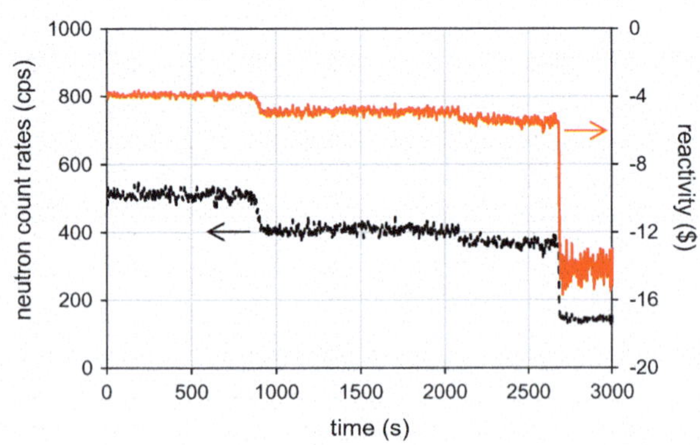

Fig. 10.1 Example of reactivity estimation using the inverse kinetics method

10.3.1.1 Simplest Reactivity Estimator (SRE)

Although the point kinetics parameters (Λ and β_{eff}) are necessary in the basic principle of the measurement, these parameters may be unknown in advance for the target core. To estimate the reactivity $\rho(t)/\beta_{\text{eff}}$ in dollar units for such a case, an improved method of the Simplest Reactivity Estimator (SRE) [11], which applies the prompt jump approximation, is useful. The principle of measurement for the improved method is explained below. First, both sides of the point kinetics Eq. (10.25) are divided by $\beta_{\text{eff}}/\Lambda$ which corresponds to the prompt neutron decay constant α_{c} at the critical state. Similarly, both sides of Eq. (10.26) are multiplied by $(\lambda_i\Lambda)/(a_i\beta_{\text{eff}})$. Then, by applying the prompt jump approximation ($dR/dt \approx 0$), the following transformed equations are obtained:

$$0 = \left(\frac{\rho(t)}{\beta_{\text{eff}}} - 1\right)R(t) + \overline{D}(t) + Q, \tag{10.32}$$

$$\frac{dD_i}{dt} = -\lambda_i D_i(t) + \lambda_i R(t), \tag{10.33}$$

$$D_i(t) \equiv \frac{\lambda_i \Lambda C_i(t)}{a_i \beta_{\text{eff}}}, \tag{10.34}$$

$$\overline{D}(t) \equiv \sum_i a_i D_i(t), \tag{10.35}$$

$$Q \equiv \frac{\Lambda S}{\beta_{\text{eff}}}. \tag{10.36}$$

Based on Eqs. (10.32) through (10.36), the following procedure enables us to successively estimate the subcriticality in dollar units without explicitly using the point kinetics parameters from the time series data of neutron count rate, which are continuously measured during a certain counting gate width Δt, R_n ($n = 0, 1, 2, \cdots$):

(1) Assume that the initial condition ($n = 0$) is a steady state ($dD_i/dt = 0$) and the initial value of $\rho_{\text{ref}}/\beta_{\text{eff}}$ is already known. Based on Eqs. (10.32) and (10.33), the following initial values of the ith delayed neutron term $D_{i,0}$ and the effective neutron source Q defined by Eqs. (10.34) and (10.36) are set depending on the initial steady-state neutron count rate R_0, respectively:

$$D_{i,0} = R_0, \tag{10.37}$$

$$Q = -\frac{\rho_{\text{ref}}}{\beta_{\text{eff}}}R_0. \tag{10.38}$$

(2) The numerical solution of $D_{i,n}$ can be successively calculated using the following recurrence Eq. (10.39) with the delayed neutron parameters (the decay constant

λ_i and the relative abundance a_i for the ith precursor group [12]). Note that Eq. (10.39) can be obtained by analytically solving the first-order differential equation of Eq. (10.33) based on the linear approximation for the time variation in the neutron count rate $R(t)$ during the time width Δt.

$$D_{i,n} = D_{i,n-1}\mathrm{e}^{-\lambda_i \Delta t} + \left(\frac{1-\mathrm{e}^{-\lambda_i \Delta t}}{\lambda_i \Delta t} - \mathrm{e}^{-\lambda_i \Delta t}\right) R_{n-1} + \left(1 - \frac{1-\mathrm{e}^{-\lambda_i \Delta t}}{\lambda_i \Delta t}\right) R_n. \tag{10.39}$$

(3) Based on Eq. (10.32), the reactivity in dollar units $\rho_n/\beta_{\mathrm{eff}}$ at the nth time step can be inversely estimated using the measured neutron count rate R_n and the numerical results of the average value of the delayed neutron terms $\overline{D}_n = \sum_i a_i D_{i,n}$ as follows:

$$\frac{\rho_n}{\beta_{\mathrm{eff}}} = 1 - \frac{\overline{D}_n + Q}{R_n}. \tag{10.40}$$

However, similar to the neutron source multiplication method, note that the principle of the inverse kinetics method also relies on the fundamental mode approximation to derive the point kinetics equation. Therefore, if the spatial distribution of the neutron flux in the target state changes significantly from the reference state used to calibrate the effective source strength, the excitation of the spatial higher-order mode components of neutron flux will result in a large systematic error in the reactivity estimated by the inverse kinetics method. In addition, if there is signal noise in the time series data of the count rate R_n due to some kind of reason (e.g., an electronic circuit for the neutron measurement), the outlier will appear in the reactivity estimated by the inverse kinetics method. In such a case, a noise filtering method (e.g., the median filter [13]) is helpful for robust estimation against the outlier in the measured R_n.

10.3.1.2 Least-Squares Inverse Kinetics Method

As can be seen from the above explanation, the effective external neutron source strength Q should be appropriately calibrated to estimate the reactivity in dollar units $\rho_n/\beta_{\mathrm{eff}}$ by the inverse kinetics method. For a critical core without the external neutron source, setting $Q = 0$ is sufficient. On the other hand, if the target system is a source-driven subcritical core and the initial subcriticality $-\rho_{\mathrm{ref}}/\beta_{\mathrm{eff}}$ is known, the value of Q can be calibrated based on Eq. (10.38) using the initial steady-state count rate R_0.

Alternatively, if the time series data of the neutron count rate R_n after the stepwise reactivity transient can be measured with sufficiently small statistical uncertainty, the source strength Q can be estimated by the least-squares inverse kinetics method [14]. For example, let us assume that the reactivity of the target core is quickly changed from the initial value to a certain value of $\rho(t) = \rho_{\mathrm{after}}$ and that the reactivity ρ_{after} is kept constant after the transient. At the mth time step after such the reactivity transient, the calculated average value $\overline{D}_m = \sum_i a_i D_{i,m}$ and the measured value of the count rate R_m are plotted on the coordinates $(\overline{D}_m, R_m)$. Then, Q can be estimated by the least-squares fitting of Eq. (10.41) which is a transformed expression of Eq. (10.32):

$$R_m = (\overline{D}_m + Q)/\left(1 - \frac{\rho_{\text{after}}}{\beta_{\text{eff}}}\right). \tag{10.41}$$

10.3.2 Measurement Procedure

In order to measure the subcriticality using the inverse kinetics method, a signal processing circuit system should be configured for the continuous measurement of the time series data of the neutron count rate, e.g., the list mode data measurement using a digital MCA as explained in Sect. 9.2. For a target core in a critical state or in a subcritical state with an external neutron source, where the initial condition is the steady state, the measurement procedure using the inverse kinetics method is described below.

(1) If the target system is a critical core, the criticality is achieved by adjusting the control rod positions. In this case, the initial condition of the reactivity is $\rho_{\text{ref}} = 0$.

- Alternatively, if the target system is a source-driven subcritical system, the initial subcritical core is configured by inserting and/or withdrawing the six control rods (three control rods and three safety rods) in an arbitrary pattern, and an external neutron source is inserted into the target core. In this case, the initial condition of the reactivity in dollar units $\rho_{\text{ref}}/\beta_{\text{eff}}$ can be obtained based on the excess reactivity and the control rod worth in advance through the stable period method and control rod calibration experiments, respectively.

(2) The initial value of the constant count rate R_0 is measured by a neutron detector (e.g., BF_3 or ^3He proportional counter) installed in the steady-state target core. For example, the neutron counts are measured for a sufficiently long time, and R_0 is estimated from the time average of the neutron counts.
(3) The effective external neutron source strength Q is calibrated using Eq. (10.38).
(4) The time series data of the neutron count rate R_n are continuously measured per a certain time width Δt, which is appropriately set according to the conditions, e.g., $\Delta t = 0.5$ s, etc.
(5) The reactivity of the target core is changed by inserting or withdrawing the control rod while measuring the neutron count rate R_n. Then, the information on the insertion or withdrawal time of the control rod is also recorded.

(6) The reactivity in dollar units $\rho_n/\beta_{\text{eff}}$ is inversely estimated based on Eqs. (10.39) and (10.40) with the measured count rate R_n. If the effective delayed neutron fraction β_{eff} is known for the target core, the reactivity in dollar units can be converted to the reactivity ρ_n.

10.3.3 Discussions

Let us discuss the measurement results of reactivity $\rho(t)$ by the inverse kinetics method, from the following viewpoints:

(1) Discuss the relationship between the time dependence of the measured count rate $R(t)$ (e.g., magnitude and sign for the slope) and the reactivity $\rho(t)$ using the inverse kinetics method. For example, how is $\rho(t)$ inversely estimated for the following change of $R(t)$?

 (a) $R(t)$ is constant.
 (b) $R(t)$ increases with time.
 (c) $R(t)$ decreases with time.

(2) Discuss how to reduce the statistical uncertainty of the reactivity $\rho(t)$ using the inverse kinetics method.
(3) What is the difference in the estimated reactivity $\rho(t)$ between the inverse kinetics method and the neutron source multiplication method described in Sect. 10.2? For example, discuss how the differences in $\rho(t)$ between these methods are (a) during the transient state of the count rate due to the control rod operation and (b) at steady state after sufficient time has elapsed following the control rod operation.
(4) By carrying out a transient experiment to which the least-squares inverse kinetics method is applicable, estimate the effective external neutron source strength Q and the reactivity in dollar units $\rho_{\text{after}}/\beta_{\text{eff}}$ after the transient, based on Eq. (10.41). For these results of Q and $\rho_{\text{after}}/\beta_{\text{eff}}$ estimated by the least-squares inverse kinetics method, compare them with the source strength Q calibrated by Eq. (10.38) using the reference subcriticality (e.g., ρ_{ref} evaluated by the excess reactivity and the control rod worth) and the reactivity estimated by the inverse kinetics method with such calibrated Q. Then, discuss the reasons for the differences between them, if any.

10.4 Pulsed Neutron Source Method

10.4.1 Principle of Measurement

A charged particle accelerator is utilized to generate a pulsed neutron source, e.g., a deuterium–tritium (D-T) neutron source can be generated by irradiating a deuterium (^{2}H) ion beam onto a ^{3}H target. Thanks to such a pulsed neutron source, the integral experiment can be carried out to measure core characteristic parameters (e.g., prompt neutron decay constant α and subcriticality in dollar units $-\rho/\beta_{\text{eff}}$) by analyzing the time variation in the neutron count rate after injecting the pulsed neutron source into a target system. This experimental technique is known as the "pulsed neutron source method," which is one of the kinetic methods. In a typical experiment of the pulsed neutron source method, a delta function-like neutron source is periodically injected into the target core, whose subcriticality $-\rho$ is kept constant.

As an example, let us assume that a single shot of pulsed fast neutrons is injected into a target subcritical system ($k_{\text{eff}} < 1$) to qualitatively explain the time variation of the thermal neutron flux in the target system. Figure 10.2 shows a typical example of the thermal neutron flux change after injecting the pulsed fast neutrons.

In Fig. 10.2, Region I is the time region where the injected fast neutrons diffuse and are thermalized. Region II is the time region where the energy spectrum of the neutron flux reaches the equilibrium state and the higher-order mode components in the spatial distribution of the neutron flux gradually decrease. Region III is the time region where the thermal neutron flux exponentially decreases with respect to the time by a specific decay constant. This decay constant due to the prompt neutron is known as the "prompt neutron decay constant." In Region IV, the thermal neutron flux is low and decreases more slowly than in Region III. The time variation of the thermal neutron flux in Region IV is mainly due to the delayed neutron component. The order of the decay constant roughly corresponds to the average of the decay constants for the delayed neutron precursors, e.g., $\overline{\lambda} \approx 1/\left(\sum_{i=1}^{6} (a_i/\lambda_i)\right)$. The amplitude of the

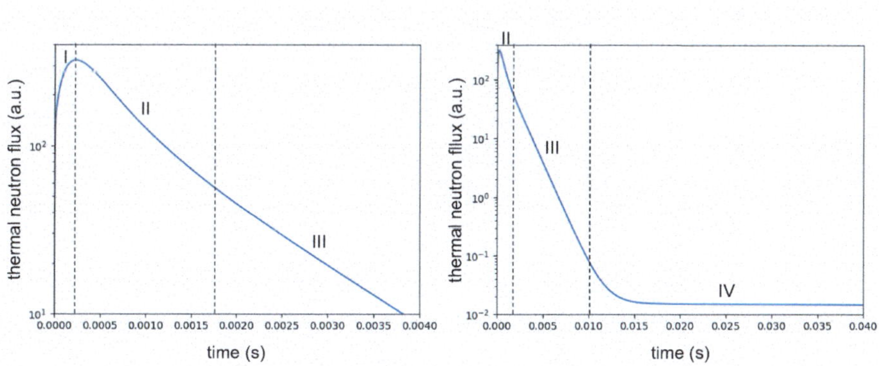

Fig. 10.2 Time variation example of thermal neutron flux after injecting pulsed fast neutrons

delayed neutron component is inversely proportional to the subcriticality in dollar units in the target system.

The following explanation focuses on the time variation of the thermal neutron flux in Regions III and IV to derive the theoretical relationship between the subcriticality and the exponential decrease of the neutron count rate measured by a thermal neutron detector. For simplicity, the fundamental mode approximation is applied in the same way as the inverse kinetics method described in Sect. 10.3. Then, the time variation of the neutron count rate $R(t)$ after injecting a pulsed neutron source at the time $t = 0$ is described by the following point kinetics equation:

$$\frac{dR}{dt} = \frac{\rho - \beta_{\mathrm{eff}}}{\Lambda} R(t) + \sum_{i=1}^{6} \lambda_i C_i(t) + S\delta(t), \tag{10.42}$$

$$\frac{dC_i}{dt} = -\lambda_i C_i(t) + \frac{a_i \beta_{\mathrm{eff}}}{\Lambda} R(t), \tag{10.43}$$

where $\delta(t)$ represents the delta function and the count rate $R(t)$ and the delayed neutron precursor density term $C_i(t)$ are zero before the injection of the pulsed neutron source. Since the target core is in a subcritical state ($\rho < 0$), the family of neutrons due to the fission chain reaction originating from the ancestor of the pulsed neutron gradually terminates so that the limit values of $R(t)$ and $C_i(t)$ at $t \to \infty$ are zero.

For further theoretical derivation, let us decompose $R(t)$ into the prompt and delayed neutron components of $R_{\mathrm{p}}(t)$ and $R_{\mathrm{d}}(t)$, respectively, as follows:

$$R(t) = R_{\mathrm{p}}(t) + R_{\mathrm{d}}(t). \tag{10.44}$$

The two components $R_{\mathrm{p}}(t)$ and $R_{\mathrm{d}}(t)$ follow the time differential equations shown in Eqs. (10.45) and (10.46), respectively. The sum of Eqs. (10.45) and (10.46) corresponds to the original point kinetics Eq. (10.42).

$$\frac{dR_{\mathrm{p}}}{dt} = -\alpha R_{\mathrm{p}}(t) + S\delta(t), \tag{10.45}$$

$$\frac{dR_{\mathrm{d}}}{dt} = -\alpha R_{\mathrm{d}}(t) + \sum_{i=1}^{6} \lambda_i C_i(t), \tag{10.46}$$

$$\alpha = \frac{\beta_{\mathrm{eff}} - \rho}{\Lambda}, \tag{10.47}$$

where α represents the prompt neutron decay constant and is proportional to the negative reactivity due to prompt neutrons only, i.e., $\beta_{\mathrm{eff}} - \rho$. The analytical solution of the prompt neutron component $R_{\mathrm{p}}(t)$ can be obtained by solving the time differential Eq. (10.45) with the initial condition of $R_{\mathrm{p}}(0) = 0$:

$$R_{\mathrm{p}}(t) = S\mathrm{e}^{-\alpha t}. \tag{10.48}$$

For the delayed neutron component $R_{\mathrm{d}}(t)$, instead of rigorously solving the analytical solution, let us consider the total count A_{d} over the time range of $0 \leq t \leq \infty$:

$$A_{\mathrm{d}} \equiv \int_0^\infty R_{\mathrm{d}}(t)dt. \tag{10.49}$$

Since $R_{\mathrm{d}}(t)$ and $C_i(t)$ are zero at $t = 0$ and ∞, the following relationship is obtained by integrating both sides of Eq. (10.46) over $0 \leq t \leq \infty$:

$$\alpha A_{\mathrm{d}} = \sum_{i=1}^{6} \lambda_i \int_0^\infty C_i(t)dt. \tag{10.50}$$

Similarly, the following equation can be obtained by integrating both sides of Eq. (10.43) over $0 \leq t \leq \infty$ and summing over all delayed neutron precursor groups i:

$$\sum_{i=1}^{6} \lambda_i \int_0^\infty C_i(t)dt = \frac{\beta_{\mathrm{eff}}}{\Lambda}\left(A_{\mathrm{p}} + A_{\mathrm{d}}\right), \tag{10.51}$$

$$A_{\mathrm{p}} \equiv \int_0^\infty R_{\mathrm{p}}(t)dt = \frac{S}{\alpha}, \tag{10.52}$$

where A_{p} denotes the total count of the prompt neutron component. By substituting Eq. (10.51) into Eq. (10.50) and rearranging using Eqs. (10.47) and (10.52), the total count of the delayed neutron component A_{d} can be derived as follows:

$$A_{\mathrm{d}} = \frac{\beta_{\mathrm{eff}}}{-\rho}A_{\mathrm{p}} - \frac{\beta_{\mathrm{eff}}}{-\rho}\frac{S}{\alpha}. \tag{10.53}$$

In an actual experiment using the pulsed neutron source method, the delta function-like pulsed neutron sources are continuously injected per a certain constant period τ to improve the statistical precision of the count rate. When one shot of the pulsed neutrons is injected into the target core at each time of $t = -j\tau$ $(j = 0, 1, 2, \cdots)$, the measured count rate $R_\tau(t)$ during $0 \leq t < \tau$ can be expressed by the following sum of $R(t + j\tau)$ due to each of the pulsed neutron shots:

$$R_\tau(t) = \sum_{j=0}^{\infty} R(t + j\tau) = \sum_{j=0}^{\infty}\left(R_{\mathrm{p}}(t + j\tau) + R_{\mathrm{d}}(t + j\tau)\right). \tag{10.54}$$

In Eq. (10.54), using Eq. (10.48), the count rate due to the prompt neutron component during $0 \le t < \tau$ can be solved as follows:

$$\sum_{j=0}^{\infty} R_{\mathrm{p}}(t + j\tau) = S\mathrm{e}^{-\alpha t} \sum_{j=0}^{\infty} \mathrm{e}^{-j\alpha\tau} = \frac{S\mathrm{e}^{-\alpha t}}{1 - \mathrm{e}^{-\alpha\tau}}. \tag{10.55}$$

On the other hand, if the experimental condition of the pulsed neutron period τ is much shorter than the average lifetime of the delayed neutron precursor (i.e., the inverse of the average decay constant), the count rate due to the delayed neutron component during $0 \le t < \tau$ can be regarded as a constant value within the time range. Therefore, using the time integral over the time range of $0 \le t < \tau$, the constant count rate can be approximately obtained as follows:

$$\sum_{j=0}^{\infty} R_{\mathrm{d}}(t + j\tau) \approx \sum_{j=0}^{\infty} \frac{1}{\tau} \int_{0}^{\tau} R_{\mathrm{d}}(t + j\tau) dt$$

$$= \frac{1}{\tau} \int_{0}^{\infty} R_{\mathrm{d}}(t) dt = \frac{A_{\mathrm{d}}}{\tau} = \frac{S}{\alpha\tau} \frac{\beta_{\mathrm{eff}}}{-\rho}. \tag{10.56}$$

Consequently, when the pulsed neutron sources are continuously injected with the constant period, the time variation of the count rate $R_\tau(t)$ during $0 \le t < \tau$ can be expressed by summing the exponential decay term of the prompt neutron component and the constant term of the delayed neutron component:

$$R_\tau(t) = R_0 \mathrm{e}^{-\alpha t} + R_{\mathrm{BG}}, \tag{10.57}$$

$$R_0 = \frac{S}{1 - \mathrm{e}^{-\alpha\tau}}, \tag{10.58}$$

$$R_{\mathrm{BG}} = \frac{S}{\alpha\tau} \frac{\beta_{\mathrm{eff}}}{-\rho}. \tag{10.59}$$

Figure 10.3 shows a typical example of the count rate $R_\tau(t)$ measured by the pulsed neutron source method. Based on Eq. (10.57) derived above, the following two measurement principles of the subcriticality measurement method are explained in the next sections: the Simmons-King method [5] and the area ratio method [4].

Note, however, that the principle of the pulsed neutron source method is also based on the point kinetics approximation which applies the fundamental mode approximation similar to the inverse kinetics method. For example, when the measurement is carried out by installing neutron detectors at multiple locations in the target subcritical core, the contribution of the higher-order mode component to the time variation of the count rate $R_\tau(t)$ depends on the spatial relationship between the core, the pulsed neutron source, and the neutron detector, as shown in Fig. 10.4 [15, 16].

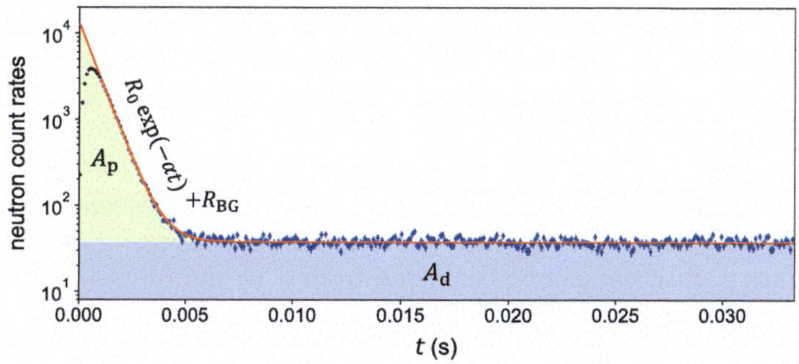

Fig. 10.3 Example of neutron count rate measured by the pulsed neutron source method

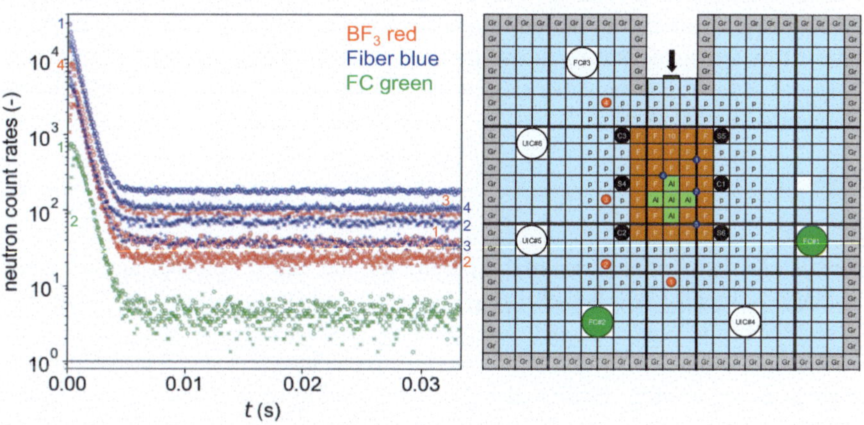

Fig. 10.4 Differences in experimental results of the pulsed neutron source method depending on the neutron detector position in the KUCA-A core

Namely, the measurement results of the Simmons-King and area ratio methods (i.e., the prompt neutron decay constant α and the area ratio A_p/A_d) are spatially dependent due to the detector position. In addition, as the subcriticality in the target core becomes deeper, the measurement result of the pulsed neutron source method may have a larger systematic bias because the larger neutron decay constant of the fundamental mode component makes it more difficult to extract only the fundamental mode component.

10.4.1.1 Simmons-King Method

As explained in Figs. 10.2 and 10.4, the higher-order mode component is included in the time variation of the neutron count rate $R_\tau(t)$ just after injecting the pulsed neutron source. By excluding the time range of $0 \leq t < t_{mask}$ where the higher-order modes exist, only the time variation of $R_\tau(t)$ during $t_{\mathrm{mask}} \le t < \tau$ is focused in order to extract as much of the exponential decay of the fundamental mode component as possible. Namely, the least-squares fitting method with Eq. (10.57) is applied only to the appropriate time range without the higher-order modes, to estimate the three fitting parameters α, R_0, and R_{BG}. Here, the fitting result of α corresponds to the prompt neutron decay constant of the fundamental mode component. Therefore, if the point kinetics parameters Λ and β_{eff} are known in advance for the target core, then the subcriticality $-\rho$ can be converted from α by the following relationship based on Eq. (10.47):

$$-\rho = \alpha\Lambda - \beta_{\mathrm{eff}}. \tag{10.60}$$

On the other hand, if Λ and β_{eff} are unknown for the target core, a different way is used to convert from the measured α into the subcriticality. First, the value of $\beta_{\mathrm{eff}}/\Lambda$ can be estimated by measuring the prompt neutron decay constant α_{ref} for a reference subcritical state where the subcriticality in dollar units $-\rho_{\mathrm{ref}}/\beta_{\mathrm{eff}}$ is already known from another measurement:

$$\alpha_{\mathrm{c}} = \frac{\beta_{\mathrm{eff}}}{\Lambda} = \frac{\alpha_{\mathrm{ref}}}{1 - \rho_{\mathrm{ref}}/\beta_{\mathrm{eff}}}, \tag{10.61}$$

where $\beta_{\mathrm{eff}}/\Lambda$ is a core characteristic parameter and called "prompt neutron decay constant at criticality α_{c}" because $\beta_{\mathrm{eff}}/\Lambda$ corresponds to the specific α value when the reference state is near the delayed criticality condition, $-\rho_{\mathrm{ref}}/\beta_{\mathrm{eff}} \approx 0$. In actual measurement, since it is not easy to directly measure α_{c} at the just critical state, the value of α_{c} is often estimated by the limit value of α when $-\rho \to 0$, e.g., by extrapolating the α values for different subcritical states of which subcriticality values can be evaluated by the excess reactivity and the control rod worth.

Consequently, if the α_{c} value can be estimated and the neutron generation time Λ can be regarded as a constant value for the target core in a near-critical state, the subcriticality in dollar units $-\rho_{\mathrm{target}}/\beta_{\mathrm{eff}}$ can be obtained by the relative difference between the measured prompt neutron decay constant of α_{target} and α_{c}:

$$\frac{-\rho_{\mathrm{target}}}{\beta_{\mathrm{eff}}} = \frac{\alpha_{\mathrm{target}}}{\alpha_{\mathrm{c}}} - 1. \tag{10.62}$$

As explained above, the subcriticality measurement method based on the prompt neutron decay constant is called the Simmons-King method [5].

10.4.1.2 Area Ratio Method

The Simmons-King method is a subcriticality measurement method that focuses on the fundamental mode of the prompt neutron decay constant. Sjöstrand proposed another pulsed neutron source method [4], which focuses on the ratio of the following two neutron counts: the exponential decay component due to the prompt neutrons and the background component of the neutron count rate due to delayed neutrons by periodically injecting the pulsed neutrons. For example, as shown in Fig. 10.3, let us consider the total neutron counts integrated over the time interval of the pulsed neutron period ($0 \leq t < \tau$) for the prompt neutron component $R_0 \exp(-\alpha t)$ and the delayed neutron component R_{BG}, respectively, in the measured count rate $R_\tau(t)$. As explained in the derivation process of Eq. (10.57), these two total counts are equivalent to A_p and A_d, which represent the integrals of the count rates $R_\mathrm{p}(t)$ and $R_\mathrm{p}(t)$, respectively, due to a shot of the pulsed neutron source over $0 \leq t \leq \infty$. Thus, using Eq. (10.53), the following relationship can be derived for the ratio of the prompt and delayed neutron components $A_\mathrm{p}/A_\mathrm{d}$:

$$\frac{A_\mathrm{p}}{A_\mathrm{d}} = \frac{\int_0^\infty R_\mathrm{p}(t)dt}{\int_0^\infty R_\mathrm{d}(t)dt} = \frac{\int_0^\tau R_0 e^{-\alpha t}dt}{\int_0^\tau R_{\mathrm{BG}}dt} = \frac{-\rho}{\beta_{\mathrm{eff}}}. \tag{10.63}$$

Alternatively, using a heuristic method based on the reactor physics, Eq. (10.63) can be derived as follows. The total neutron counts of both components $A_\mathrm{p} + A_\mathrm{d}$ is inversely proportional to $(1 - k_{\mathrm{eff}})$ because a pulsed neutron source S is amplified by infinite geometric series with the effective multiplication factor k_{eff}:

$$A_\mathrm{p} + A_\mathrm{d} \propto S\left(1 + k_{\mathrm{eff}} + k_{\mathrm{eff}}^2 + \cdots\right) = \frac{S}{1 - k_{\mathrm{eff}}}. \tag{10.64}$$

Similarly, the total neutron count of the prompt neutron component A_p is inversely proportional to $\{1 - (1 - \beta_{\mathrm{eff}})k_{\mathrm{eff}}\}$ by considering the prompt neutron multiplication factor $(1 - \beta_{\mathrm{eff}})k_{\mathrm{eff}}$ without the delayed neutrons:

$$A_\mathrm{p} \propto \frac{S}{1 - (1 - \beta_{\mathrm{eff}})k_{\mathrm{eff}}}. \tag{10.65}$$

By taking the ratio of both sides of Eqs. (10.64) and (10.65), the following relationship can be obtained, which is equivalent to Eq. (10.63):

$$\frac{A_\mathrm{p} + A_\mathrm{d}}{A_\mathrm{p}} = 1 + \frac{A_\mathrm{d}}{A_\mathrm{p}} = \frac{1 - (1 - \beta_{\mathrm{eff}})k_{\mathrm{eff}}}{1 - k_{\mathrm{eff}}} = 1 - \frac{\beta_{\mathrm{eff}}}{\rho}. \tag{10.66}$$

As explained above, since the absolute value of the subcriticality in dollar units can be measured by analyzing the ratio of the total neutron counts (or the areas of the time integrals of the measured count rates) of the prompt to the delayed neutron components $A_\mathrm{p}/A_\mathrm{d}$, this measurement method is called the "area ratio method" [4]. The area ratio method is one of the few methods to measure the absolute value of the subcriticality from the experimental result for the target system only, without any calibration measurement in a reference system with the known subcriticality. In an actual experiment of the area ratio method, the following two methodologies can be utilized to measure A_p and A_d:

(1) Direct measurement for A_p and A_d: For example, the background component of the count rate R_{BG} is measured within the time range of $t_{cut} \leq t < \tau$, where the delayed neutron component is dominant by setting t_{cut} so that the prompt neutron component decreases negligibly, e.g., $t_{cut} > 5/\alpha$ since $e^{-5} < 0.01$. Then, A_d can be estimated by $R_{BG}\tau$, i.e., the product of the background count rate R_{BG} and the pulsed neutron period τ. In addition, the total count of $A_p + A_d$ can be directly measured within the time range of the pulsed neutron period, $0 \leq t < \tau$. Thus, A_p can be obtained by subtracting A_d from $A_p + A_d$.
(2) The least-squares fitting method for $R_\tau(t)$: The time variation of the count rate $R_\tau(t)$ is successively measured with a certain counting gate width Δt. Then, the parameters α, R_0, and R_{BG} can be estimated by fitting Eq. (10.57) for the measured $R_\tau(t)$ during the appropriate time interval of $t_{mask} \leq t < \tau$ where the exponential decay of the fundamental mode component is dominant after the decay of the higher-order mode components. Based on Eq. (10.57), the areas of A_p and A_d can be obtained by integrating the fitting result of $R_\tau(t)$ over the time range of $0 \leq t < \tau$. As illustrated in Fig. 10.3, this method is also called the "extrapolated area ratio method" [17], because the area ratio is evaluated by extrapolating the prompt neutron component in the time domain that includes the higher-order mode effect, to extract the fundamental mode.

10.4.1.3 Dynamic Mode Decomposition (DMD)

When multiple detectors are installed in a target core, various techniques [18, 19] have been proposed to robustly estimate the prompt neutron decay constant α and the area ratio A_p/A_d corresponding to the fundamental mode component. Recently, "dynamic mode decomposition (DMD)" [15, 16, 20] has been applied to the pulsed neutron source method as a data-driven analysis method. An overview of the analysis method using DMD is explained below.

(1) M neutron detectors are installed at different positions in the target core to observe the different higher-order mode effects in the measured neutron count rates at these detector positions. The experiment with the pulsed neutron source method is carried out to obtain the time series data of neutron counts $R_{m,n}$ with N time steps of a certain interval Δt, where $1 \leq m \leq M$, $1 \leq n \leq N$, and $M < N$. To easily separate the background components due to the delayed neutrons, a stationary signal $R_{M+1,n} = 1$ is additionally considered. Then, the data matrix $\mathbf{X} \equiv \left(\vec{R}_1\ \vec{R}_2\ \cdots\ \vec{R}_N\right) \in \mathbb{R}^{(M+1)\times N}$ is prepared by arranging the elements $R_{m,n}$, where $\vec{R}_n = \left(R_{1,n}\ R_{2,n}\ \cdots\ R_{M+1,n}\right)^T$ is the $(M+1)$ dimensional column vector at the nth time step.
(2) The following two slicing matrices are obtained by taking the first or the second through the $(N-1)$th or the Nth time series data from the original matrix $\mathbf{X}$, i.e., $\mathbf{X}_{1:N-1} \equiv \left(\vec{R}_1\ \vec{R}_2\ \cdots\ \vec{R}_{N-1}\right)$ and $\mathbf{X}_{2:N} \equiv \left(\vec{R}_2\ \vec{R}_3\ \cdots\ \vec{R}_N\right)$.
(3) The relationship between $\mathbf{X}_{1:N-1}$ and $\mathbf{X}_{2:N}$ is assumed using the time evolution matrix $\mathbf{A} \in \mathbb{R}^{(M+1)\times(M+1)}$ as follows:

$$\mathbf{X}_{2:N} = \mathbf{A}\mathbf{X}_{1:N-1}. \tag{10.67}$$

(4) The slicing matrix $\mathbf{X}_{1:N-1}$ can be expressed by the following singular value decomposition:

$$\mathbf{X}_{1:N-1} = \mathbf{U}\boldsymbol{\Sigma}\mathbf{V}^*, \tag{10.68}$$

where $\mathbf{U} \in \mathbb{R}^{(M+1)\times r}$ and $\mathbf{V} \in \mathbb{R}^{(N-1)\times r}$ are semi-unitary matrices consisting of left and right singular vectors; $\Sigma \in \mathbb{R}^{r\times r}$ is the diagonal matrix whose elements correspond to the singular values; $r \leq (M+1)$ is the rank of the matrix $\mathbf{X}_{1:N-1}$; and the superscript * denotes the conjugate transpose.

(5) To estimate the eigenvalue of the time evolution matrix $\mathbf{A}$, the projected matrix $\tilde{\mathbf{A}} \in \mathbb{R}^{r\times r}$ is calculated using $\mathbf{U}$ as follows:

$$\tilde{\mathbf{A}} = \mathbf{U}^*\mathbf{A}\mathbf{U} = \mathbf{U}^*\mathbf{X}_{2:N}\mathbf{V}\boldsymbol{\Sigma}^{-1}, \tag{10.69}$$

where Σ^{-1} represents the inverse matrix of Σ.

(6) By the eigenvalue decomposition for $\tilde{\mathbf{A}}$, the eigenvalues μ_i and the eigenvectors $\vec{w}_i$ are obtained $(1 \leq i \leq r)$, where the order i of μ_i and $\vec{w}_i$ is arranged in descending order of the magnitude of $\ln(|\mu_i|)/\Delta t$.

(7) The matrix of the eigenvectors for the time evolution matrix $\mathbf{A}$, $\Phi = \left(\vec{\phi}_1\ \vec{\phi}_2 \cdots \vec{\phi}_r\right) \in \mathbb{R}^{(M+1)\times r}$, can be obtained by the following matrix operation:

$$\Phi = \mathbf{X}_{2:N}\mathbf{V}\boldsymbol{\Sigma}^{-1}\mathbf{W}\mathbf{diag}(1/\mu_i), \tag{10.70}$$

where $\mathbf{W} = \left(\vec{w}_1\ \vec{w}_2 \cdots \vec{w}_r\right) \in \mathbb{R}^{r\times r}$ is the matrix of the eigenvectors for the projected matrix $\tilde{\mathbf{A}}$; and $\mathbf{diag}(1/\mu_i)$ is the diagonal matrix whose element corresponds to the inverse of the eigenvalue $1/\mu_i$.

(8) Using the estimated μ_i and $\vec{\phi}_i$, the column vector of the measured neutron count rate $\vec{R}(t)$ at the time t can be expanded as follows:

$$\vec{R}(t) = \sum_{i=1}^{r} C_i\vec{\phi}_i \exp(\omega_i(t - t_s)), \tag{10.71}$$

$$\vec{C} = \left(C_1\ C_2 \cdots C_r\right)^{\mathrm{T}} = \Phi^+\vec{R}_1, \tag{10.72}$$

$$\omega_i = \frac{\ln(\mu_i)}{\Delta t}, \tag{10.73}$$

where Φ^+ is the pseudo-inverse matrix of Φ; and $\vec{R}_1$ represents the neutron count rate at the initial time t_s which corresponds to the masking time from the injection time of the pulsed neutron source $t = 0$

(9) Note that the modes of $i = 1$ and 2 in Eq. (10.71) correspond to the constant term $e^{0(t-t_s)} = 1$ (i.e., the zero decay constant $\omega_1 = 0$) and the exponential decay term $\exp(-\alpha(t - t_s))$ (i.e., the prompt neutron decay constant of the fundamental mode $\omega_2 = -\alpha$), respectively. Therefore, the higher-order mode components in Eq. (10.71) can be filtered by eliminating the terms of $i \geq 3$:

$$\vec{R}(t) \approx C_2\vec{\phi}_2 \exp(\omega_2(t - t_s)) + C_1\vec{\phi}_1. \tag{10.74}$$

(10) Finally, by applying DMD to the experimental results of the pulsed neutron sour method using the M neutron detectors, the prompt neutron decay constant α and the area ratio can be analyzed as follows:

$$\alpha_{\mathrm{DMD}} = -\omega_2 = -\frac{\ln(\mu_2)}{\Delta t}, \tag{10.75}$$

$$\left(\frac{A_{\mathrm{p}}}{A_{\mathrm{d}}}\right)_{\mathrm{DMD}} = \frac{\left\| \int_0^{\tau} C_2\vec{\phi}_2 e^{-\alpha(t'-t_s)} dt' \right\|}{\left\| \int_0^{\tau} C_1\vec{\phi}_1 dt' \right\|} = \left| \frac{C_2 e^{\alpha t_s}}{\alpha C_1 \tau} \right|. \tag{10.76}$$

10.4.2 Measurement Procedure

In order to carry out the experiment of the pulsed neutron source method, an accelerator-driven neutron source is required to periodically inject pulsed neutrons with a constant period into a target subcritical core. For example, the experiments using the pulsed neutron source were previously carried out at KUCA by utilizing the following accelerator-driven neutron sources [21]:

(1) D-T neutron source: By irradiating a tritium (^{3}H) target with a deuteron (^{2}H or d), a 14 MeV neutron is produced due to the ^{3}H(d,n)^{4}He reaction. The D-T neutron generator system consists of a duoplasmatron-type ion source, a high-voltage generator (maximum acceleration voltage 300 kV), an acceleration tube, a beam pulsing system, and a tritium target. Using the beam pulsing system, the accelerated deuterium ion beam is shaped into a pulsed shape and injected onto the tritium target to obtain the pulsed neutron source.
(2) Spallation neutron source: By irradiating a heavy metal target (lead–bismuth, tungsten, etc.) with a high-energy (100 MeV) proton beam by an FFAG (Fixed-Field Alternating Gradient) accelerator [22–24], a spallation reaction occurs. The spallation reaction can produce high-energy neutrons whose energy spectrum has a peak at ~ 2 MeV and a unique distribution from 10 to 100 MeV [25]. By injecting the pulsed proton beam onto the target, the pulsed spallation neutron source can also be generated.

In the subcriticality measurement using the pulsed neutron source method, a signal processing circuit system should be configured to continuously measure time series data of neutron count rate, e.g., the list mode data measurement using a high-speed digital multi-channel analyzer (MCA) as described in Sect. 9.2. In addition, the trigger signal to generate the pulsed neutron is also branched and measured simultaneously using another channel of the MCA. Thereby, the generation time information on each pulsed neutron source can also be measured as the list mode data.

Using the configured signal processing circuit system as described above, the prompt neutron decay constant α and the area ratio A_p/A_d can be measured as follows:

(1) By installing neutron detectors (e.g., BF_3 or ^3He proportional counter) in the target subcritical core, the signal processing circuit system is configured to carry out the pulsed neutron source method.
(2) The target subcritical system is configured by inserting and/or withdrawing six control rods (three control rods and three safety rods) in an arbitrary pattern. If possible, the neutron count rate without the pulsed neutron source is measured to confirm that the background count rate due to factors other than the pulsed neutron source is negligibly small.
(3) The accelerator operation is started to generate the pulsed neutron sources. The pulse-repetition frequency (or the pulse period τ) is appropriately set depending on the magnitude of the prompt neutron constant α in the target system (e.g., $\tau \approx 10/\alpha$). After sufficient time has passed since the pulsed neutron sources have been periodically injected, the moving average of the neutron count rate is checked to confirm the steady state of the target system.
(4) Using the digital MCA for the list mode measurement, the time series data of the trigger signal of the pulsed neutron source (or the pulsed neutron source generation time) and the neutron detection time are continuously measured. The measurement time of the pulsed neutron source method is determined according to the required statistical precision of the neutron count rate.
(5) Based on the measured list mode data, the time series data of the neutron count rate per Δt are analyzed over the time range between the jth and the next $(j+1)$th trigger signal time (i.e., $t_j \le t < t_{j+1}$). By summing (or averaging) the neutron count rates for $1 \le j \le N$ where N is the total number of the pulsed neutron shots, the time variation of the count rate $R(t)$ in the pulsed neutron source experiment can be obtained.
(6) By fitting Eq. (10.57) to the measured $R(t)$ over the time range excluding the higher-order mode component, the prompt neutron decay constant α and the area ratio A_p/A_d can be evaluated.
(7) By procedures (3) through (6) for different subcritical cores by changing the control rod pattern, α_i and $\left(A_p/A_d\right)_i$ are measured for the ith subcritical system.
(8) If the point kinetics parameters can be given, the subcriticality $-\rho$ can be estimated by Eq. (10.60) with the measurement result of α.

10.4.3 Discussions

Let us discuss the measurement results by the pulsed neutron source method, the prompt neutron decay constant α and the area ratio A_p/A_d, from the following viewpoints.

(1) Evaluate the reference value of subcriticality $-\rho_{ref,i}$ for the ith subcritical core, based on the experimental results of the excess reactivity and the control rod worth. Make a scatter plot of the measured α_i against $-\rho_{ref,i}$. Then, estimate the prompt neutron decay constant at criticality α_c by extrapolating the α value at the critical state (or $-\rho = 0$).

- Or estimate α_c using the area ratio $\left(A_p/A_d\right)_i$ as the subcriticality in dollar units in Eq. (10.61).

(2) Check how well the subcriticality obtained by (i) the Simmons-King method of Eq. (10.62) agrees with that of (ii) the area ratio method (or the extrapolated area ratio method). If there is a difference between these methods, discuss the cause of the difference by changing the masking time in the fitting process of the pulsed neutron method.

(3) Calculate the reactivity ρ by substituting $-\alpha$ for ω (i.e., $\omega = -\alpha$) in the reactivity equation of Eq. (2.62) using the point kinetics parameters (prompt neutron lifetime ℓ or neutron generation time Λ and effective delayed neutron fraction β_{eff}) and the delayed neutron parameters. Then, compare the calculated ρ with the subcriticality estimated by the Simmons-King method of Eq. (10.62). Through this comparison, discuss the relationship between the prompt neutron decay constant α and the solution for ω_j of Eq. (2.62).

(4) Based on (i) the spatial relationship between the core, the pulsed neutron source, and the neutron detector and (ii) the magnitude of the subcriticality of the target core, investigate whether the experimental results of the pulsed neutron source method (time variation of the count rate $R_\tau(t)$, prompt neutron decay constant α, and the area ratio A_p/A_d) differ depending on the neutron detector position. Then, discuss the reasons for these differences.

- By comparing the measured subcriticality with the reference value $-\rho_{ref}$ based on the excess reactivity and the control rod worth, discuss what kind of the condition for (i) the detector position and (ii) the magnitude of the subcriticality is suitable for applying the fundamental mode approximation.

10.5 Feynman-α Method

10.5.1 Principle of Measurement

Let us assume that the neutron counts are successively measured during a certain gate width T in a steady-state subcritical core with an external neutron source. Note that the thermal feedback effect is negligible because the thermal power due to the fission reaction is close to zero (i.e., zero-power). In this case, the measured neutron counts statistically fluctuate around a mean value, as shown in Fig. 10.5. The subcriticality measurement method based on the mean value of the neutron count is called the static method such as the neutron source multiplication method as explained in Sect. 10.2. The amount of the statistical fluctuation (or "reactor noise"), which corresponds to the random increase and decrease around the mean, is related to the statistical precision in the static method, and sometimes seems to be undesirable to measure the experimental result with the high precision. However, the reactor noise also includes useful information relating to the subcriticality, i.e., the temporal decay of the neutron family due to the fission chain reaction. As explained in detail later in this section, the reason is derived from the physical phenomenon that the reactor noise increases as the target subcritical core approaches the critical state, as shown in Fig. 10.5. Based on this unique physical phenomenon, the technique of measuring subcriticality by analyzing the statistical fluctuation of the neutron count is known as the "reactor noise analysis method."

Although various techniques of the reactor noise analysis methods [7, 8] have been proposed, this section explains the Feynman-α method [9]. Using the reactor noise measurement under the steady state without time variations in the subcriticality and the external neutron source strength, the Feynman-α method can be utilized to analyze the prompt neutron decay constant α. Then, the measured α value can be converted into the subcriticality $-\rho$ by the Simmons-King method [5], which is explained in Sect. 10.4 as the pulsed neutron source method. The theoretical formula of the Feynman-α method can be obtained by various derivation methods proposed so far, e.g., the Kolmogorov forward equation [26], the Pál-Bell backward equation [27, 28], and so on. In this section, the theoretical derivation of the Feynman-α method is explained using the pair detection probability, which can be derived by the heuristic method [29].

For simplicity, let us derive the theoretical formula based on the one-neutron energy group approximation for an infinite homogeneous system with an external neutron source. Here, the external neutron source is assumed to be an (α,n) source (e.g., Am-Be) or a spontaneous fission source (e.g., ^{252}Cf), which emits one or more neutrons, respectively, due to the radioactive decay. In the theoretical derivation process, the following probabilities are defined:

- S is the probability of the radioactive decay of the external neutron source per unit time. In addition, $p_s(q)$ is the probability that q neutrons are simultaneously emitted per decay.

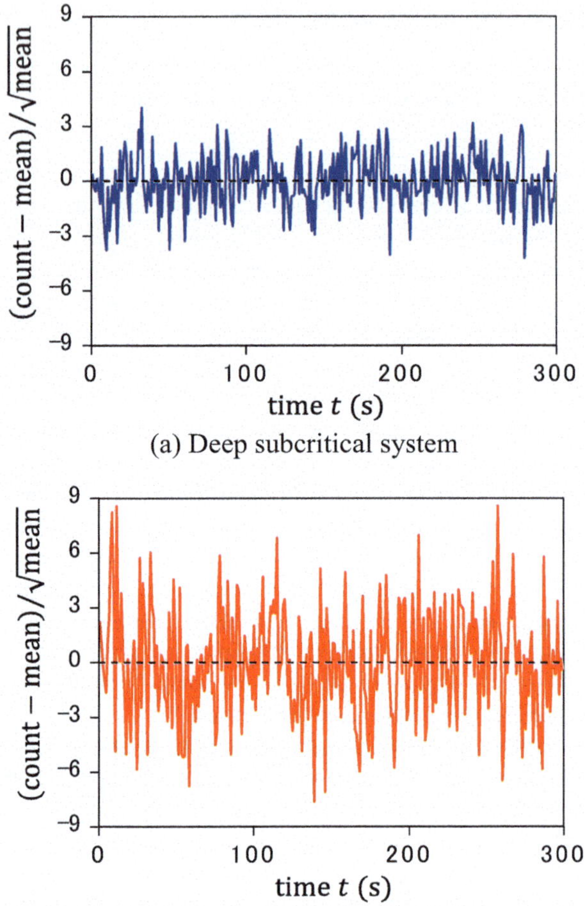

Fig. 10.5 Example of statistical fluctuations in neutron counts (reactor noise)

(a) Deep subcritical system

(b) Shallow subcritical system

- $\lambda_f = v\Sigma_f$ is the probability of a fission reaction per unit time when one neutron flies at the speed of v in the target system, where Σ_f is the macroscopic fission cross section. In addition, $p_f(\nu)$ is the probability that ν fission neutrons are simultaneously produced per fission reaction. For example, Fig. 10.6 shows the probability distribution $p_f(\nu)$ of the thermal fission of ^{235}U [30]. Using the quantities defined above, the neutron generation time Λ is approximately expressed as follows:

$$\Lambda \approx \frac{1}{\langle \nu \rangle \lambda_f}, \tag{10.77}$$

where the bracket < > means the expected value.

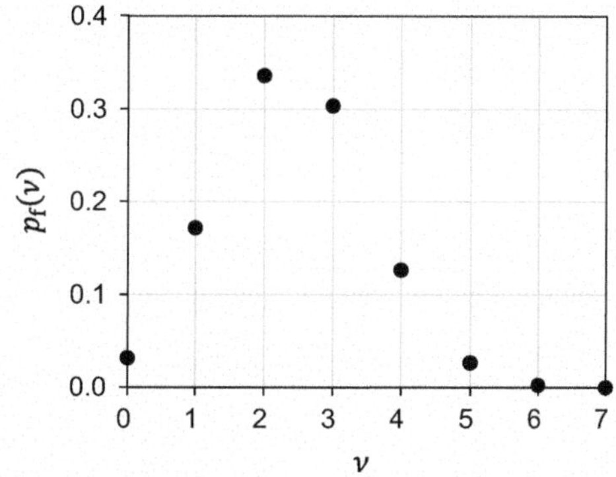

Fig. 10.6 Probability distribution of the number of prompt neutrons per thermal fission reaction of ^{235}U

- $\lambda_d = v\Sigma_d$ is the probability of a neutron detection reaction per unit time when one neutron flies in the target system, where Σ_d is the macroscopic detection cross section.

10.5.1.1 Green's Function

To derive the theoretical formula by the heuristic method, the Green's function $P_G(t|t_0)$ is utilized. Here, $P_G(t|t_0)$ represents the expected value of the descendant neutrons surviving at the time t, after one ancestor neutron is injected into the target system at the time t_0.

To solve the approximated solution $P_G(t|t_0)$, let us assume that the ancestor neutron is born like a pulsed neutron source at $t = t_0$, i.e., the delta function of $\delta(t - t_0)$ and that the prompt neutron component is dominant by focusing on the time scale of the prompt neutron lifetime. Thus, by solving the point kinetics equation of Eq. (10.78) without the delayed neutron component, the analytical solution of $P_G(t|t_0)$ can be approximated by Eq. (10.79):

$$\frac{dP_G}{dt} \approx -\alpha P_G(t|t_0) + \delta(t - t_0), \tag{10.78}$$

$$P_G(t|t_0) \approx e^{-\alpha(t-t_0)}. \tag{10.79}$$

As can be seen from Eq. (10.79), the neutron family due to the fission chain reaction originates from one ancestor neutron and decreases exponentially with respect to time. As the subcriticality of the target system becomes shallower, the neutron family persists longer. Namely, the exponential decay constant corresponds to the prompt neutron decay constant α which is proportional to the subcriticality $-\rho$.

In the theoretical derivation of $P_G(t|t_0)$, the contribution due to the delayed neutron can also be considered by adding the simultaneous time differential equations of the delayed neutron precursors to the point kinetics equation of Eq. (10.78). Although the detailed derivation process is omitted, the following analytical solution of $P_G(t|t_0)$ can be obtained using the Laplace transform:

$$P_G(t|t_0) = \Lambda \sum_{j=1}^{7} A_j e^{-\alpha_j(t-t_0)}, \tag{10.80}$$

where α_j is the jth root, which can be solved by considering the condition that the denominator of the following function $G(s)$ of Eq. (10.81) satisfies zero (i.e., the solution of the reactivity equation of Eq. (2.62), $\omega_j = -\alpha_j$, $1 \le j \le 7$); and A_j is the jth coefficient of the partial fraction decomposition for $G(s)$:

$$G(s) = \frac{1}{s\left(\Lambda + \beta_{\mathrm{eff}} \sum_{i=1}^{6} \frac{a_i}{s+\lambda_i}\right) - \rho} = \sum_{j=1}^{7} \frac{A_j}{s + \alpha_j}, \tag{10.81}$$

where $G(s)$ is known as the zero-power transfer function. Based on Eq. (10.81), the relationships of Eqs. (10.82) through (10.84) can be derived to help the theoretical derivation of the Feynman-α method as explained later:

$$G(0) = \frac{1}{-\rho} = \sum_{j=1}^{7} \frac{A_j}{\alpha_j} = \sum_{j=1}^{7} \int_0^{\infty} A_j e^{-\alpha_j t'} dt', \tag{10.82}$$

$$G(\alpha_k) = \sum_{j=1}^{7} \frac{A_j}{\alpha_j + \alpha_k} = \sum_{j=1}^{7} \int_0^{\infty} A_j e^{-(\alpha_j+\alpha_k)t'} dt', \tag{10.83}$$

$$A_j = \lim_{s \to -\alpha_j} (s + \alpha_j) G(s) = \frac{\alpha_j}{-\rho + \beta_{\mathrm{eff}} \sum_{i=1}^{6} a_i \left(\frac{\alpha_j}{\lambda_i - \alpha_j}\right)^2}. \tag{10.84}$$

10.5.1.2 Single Detection Probability

First, let us derive the "single detection probability" $P_1(t_1)dt_1$ that one neutron is detected within the time range from t_1 to $t_1 + dt_1$ when the external neutron source has been emitting neutrons since an infinite past. As shown in Fig. 10.7, the probability $P_1(t_1)dt_1$ can be obtained by taking the product of the following probabilities, followed by the summation over $0 \le q \le q_{\max}$ and the integral over the time interval of $-\infty \le t_s \le t_1$:

(1) Sdt_s is the probability that the external neutron source decay within the time range from t_s to $t_s + dt_s$.
(2) q neutrons are emitted by this external neutron source decay with the probability $p_s(q)$.

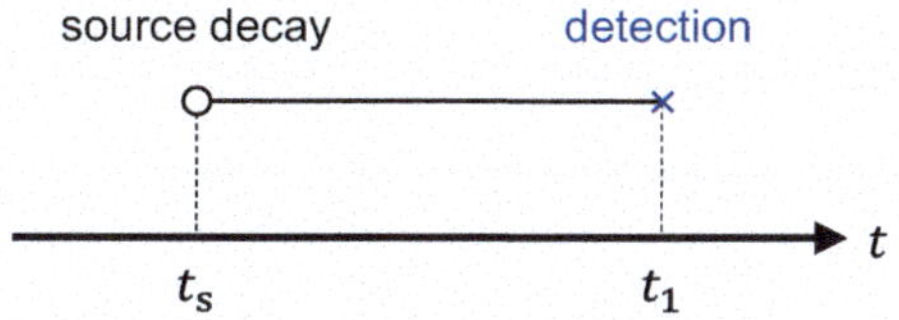

Fig. 10.7 Single detection probability

(3) The expected number of the descendant neutrons is $qP_G(t_1|t_s)$ at the time t_1.
(4) $\lambda_d dt_1$ is the probability that the descendant neutron is detected within the time range from t_1 to $t_1 + dt_1$.

$$P_1(t_1)dt_1 = \int_{-\infty}^{t_1} dt_s S \sum_{q=0}^{q_{\max}} p_s(q) q P_G(t_1|t_s)\lambda_d dt_1 = \frac{\lambda_d \langle q \rangle S \Lambda}{-\rho} dt_1,$$

$$\because \sum_{j=1}^{7} \int_{-\infty}^{t_1} A_j e^{-\alpha_j (t_1 - t_s)} dt_s = \sum_{j=1}^{7} \int_{0}^{\infty} A_j e^{-\alpha_j t'} dt' = \sum_{j=1}^{7} \frac{A_j}{\alpha_j} = G(0). \quad (10.85)$$

10.5.1.3 Pair Detection Probability

Similarly, let us derive the "pair detection probability" $P_2(t_1, t_2)dt_1 dt_2$ that one neutron is detected within the time range from t_1 to $t_1 + dt_1$ and another neutron is detected within the next time range t_2 and between $t_2 + dt_2$, as shown in Fig. 10.8. The probability $P_2(t_1, t_2)dt_1 dt_2$ is decomposed into the sum of the following three components:

$$P_2(t_1, t_2)dt_1 dt_2 = P_{2,2}(t_1, t_2)dt_1 dt_2 + P_{2,f}(t_1, t_2)dt_1 dt_2 + P_{2,s}(t_1, t_2)dt_1 dt_2. \quad (10.86)$$

First, $P_{2,2}(t_1, t_2)dt_1 dt_2$ represents the component that each of a pair of neutrons belongs to two independent neutron families with different ancestor neutrons. The probability $P_{2,2}(t_1, t_2)dt_1 dt_2$ can be obtained by the product of two single detection probabilities $P_1(t_1)dt_1$ and $P_2(t_2)dt_2$ using Eq. (10.85) as follows:

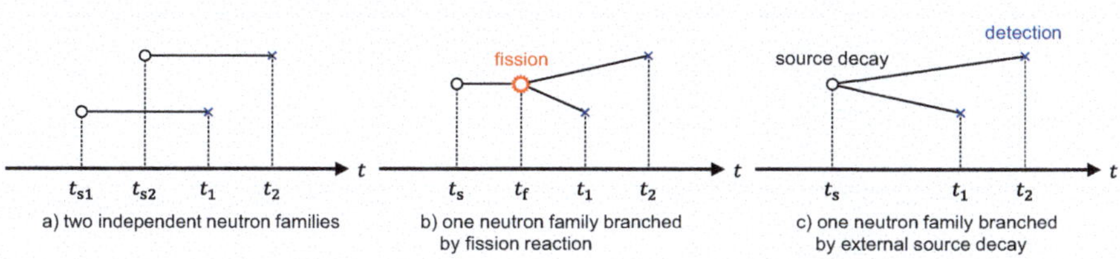

Fig. 10.8 Pair detection probability

$$P_{2,2}(t_1, t_2)dt_1dt_2 = P_1(t_1)P_1(t_2)dt_1dt_2 = \left(\frac{\lambda_d\langle q\rangle S\Lambda}{-\rho}\right)^2 dt_1dt_2. \tag{10.87}$$

Next, $P_{2,f}(t_1, t_2)dt_1dt_2$ means the component that a pair of neutrons belongs to the same neutron family and is detected after the branching process due to a fission reaction event. The probability $P_{2,f}(t_1, t_2)dt_1dt_2$ can be obtained by taking the product of the following probabilities, followed by the summation over $0 \le q \le q_{max}$ and $0 \le \nu \le \nu_{max}$, and by the integral over the time intervals of $-\infty \le t_s < t_f$ and $-\infty \le t_f \le t_1$:

(1) Sdt_s is the probability of the decay of the external neutron source.
(2) $p_s(q)$ is the probability of emitting q neutrons per the external source decay.
(3) $qP_G(t_f|t_s)$ is the expected number of descendant neutrons at the time t_f.
(4) $\lambda_f dt_f$ is the probability that the one neutron causes a fission reaction within the time range from t_f to $t_f + dt_f$.
(5) $p_f(\nu)$ is the probability of producing ν neutrons per the fission event.
(6) $\nu P_G(t_1|t_f)$ is the expected number of descendant neutrons at the time t_1.
(7) $\lambda_d dt_1$ is the probability that one descendant neutron is detected within the time range from t_1 to $t_1 + dt_1$.
(8) On the other hand, $(\nu - 1)P_G(t_2|t_f)$ is the expected number of the descendant neutrons, excluding one neutron detected in steps (6) and (7), at the time t_2 $(t_1 < t_2)$.
(9) $\lambda_d dt_2$ is the probability that another descendant neutron is detected within the time range from t_2 to $t_2 + dt_2$.

$$\begin{aligned}
P_{2,f}(t_1, t_2)dt_1dt_2 &= \int_{-\infty}^{t_1} dt_f \int_{-\infty}^{t_f} dt_s S \sum_{q=0}^{q_{max}} p_s(q)qP_G(t_f|t_s) \\
&\quad \times \lambda_f \sum_{\nu=0}^{\nu_{max}} p_f(\nu)\nu P_G(t_1|t_f)\lambda_d dt_1 (\nu - 1)P_G(t_2|t_f)\lambda_d dt_2 \\
&= \frac{\lambda_d^2\langle q\rangle S\langle\nu(\nu-1)\rangle\lambda_f\Lambda^3}{-\rho} \sum_{j=1}^{7} A_j G(\alpha_j)e^{-\alpha_j(t_2-t_1)},
\end{aligned}$$

$$\begin{aligned}
&\because \sum_{j=1}^{7}\sum_{k=1}^{7}\sum_{l=1}^{7} \int_{-\infty}^{t_f} A_l e^{-\alpha_l(t_f-t_s)}dt_s \int_{-\infty}^{t_1} A_k e^{-\alpha_k(t_1-t_f)} A_j e^{-\alpha_j(t_2-t_f)}dt_f \\
&= \sum_{j=1}^{7}\left(\sum_{k=1}^{7}\frac{A_k}{\alpha_k+\alpha_j}\right)\left(\sum_{l=1}^{7}\frac{A_l}{\alpha_l}\right)A_j e^{-\alpha_j(t_2-t_1)} \\
&= G(0)\sum_{j=1}^{7} A_j G(\alpha_j)e^{-\alpha_j(t_2-t_1)}.
\end{aligned} \tag{10.88}$$

Third, $P_{2,\mathrm{s}}(t_1, t_2)dt_1dt_2$ means the component that a pair of neutrons belongs to the same neutron family and is detected after the branching process due to the first event of the external neutron source decay. Based on the same procedure as the derivation of Eq. (10.88), the probability $P_{2,\mathrm{s}}(t_1, t_2)dt_1dt_2$ can be obtained as follows:

$$\begin{aligned} P_{2,\mathrm{s}}(t_1, t_2)dt_1dt_2 &= \int_{-\infty}^{t_1} dt_\mathrm{s} S \sum_{q=0}^{q_{\max}} p_\mathrm{s}(q) q P_\mathrm{G}(t_1|t_\mathrm{s}) \lambda_\mathrm{d} dt_1 \\ &\quad \times (q-1) P_\mathrm{G}(t_2|t_\mathrm{s}) \lambda_\mathrm{d} dt_2 \\ &= \lambda_\mathrm{d}^2 S \langle q(q-1)\rangle \Lambda^2 \sum_{j=1}^{7} A_j G(\alpha_j) \mathrm{e}^{-\alpha_j(t_2-t_1)}. \end{aligned} \tag{10.89}$$

Finally, using Eqs. (10.77) and (10.87) through (10.89), the following analytical solution for the pair detection probability $P_2(t_1, t_2)dt_1dt_2$ can be obtained:

$$P_2(t_1, t_2)dt_1dt_2 = \left\{ \left(\frac{\lambda_\mathrm{d}\langle q\rangle S\Lambda}{-\rho} \right)^2 + \frac{1}{2} \left(\frac{\lambda_\mathrm{d}\langle q\rangle S\Lambda}{-\rho} \right) \sum_{j=1}^{7} Y_{\infty,j} \alpha_j \mathrm{e}^{-\alpha_j(t_2-t_1)} \right\} dt_1dt_2, \tag{10.90}$$

$$Y_{\infty,j} \equiv 2\varepsilon \frac{\langle \nu(\nu-1)\rangle}{\langle \nu\rangle^2} \left(1 + \frac{\langle \nu\rangle\langle q(q-1)\rangle}{\langle q\rangle\langle \nu(\nu-1)\rangle} (-\rho) \right) \frac{A_j G(\alpha_j)}{\alpha_j}, \tag{10.91}$$

$$\varepsilon \equiv \frac{\lambda_\mathrm{d}}{\lambda_\mathrm{f}} = \frac{\Sigma_\mathrm{d}}{\Sigma_\mathrm{f}}. \tag{10.92}$$

As a supplementary explanation, $P_2(t_1, t_2)dt_1dt_2$ closely related to the principle of the Rossi-α method [10] and can also be utilized as the theoretical formula.

10.5.1.4 Neutron Correlation Factor Y

The single and pair detection probabilities $P_1(t_1)dt_1$ and $P_2(t_1, t_2)dt_1dt_2$ are utilized to derive the first- and second-order moments of the neutron count $C(T)$ during a certain counting gate width T. First, the first-order moment (or mean) $\langle C(T)\rangle$ can be obtained by integrating $P_1(t_1)dt_1$ over the time interval of $0 \le t_1 \le T$:

$$\langle C(T)\rangle = \int_0^T P_1(t_1)dt_1 = \frac{\lambda_\mathrm{d}\langle q\rangle S\Lambda}{-\rho} T. \tag{10.93}$$

Similarly, by integrating $P_2(t_1, t_2)dt_1dt_2$ over the time intervals of $0 \leq t_1 < t_2$ and $0 \leq t_2 \leq T$ and using Eq. (10.93), the second-order factorial moment of $C(T)$ (or the expected value of the neutron pair measured during T) can be obtained as follows:

$$\begin{aligned}\frac{\langle C(T)(C(T)-1)\rangle}{2} &= \int_0^T dt_2 \int_0^{t_2} dt_1 P_2(t_1, t_2) \\ &= \frac{1}{2}\langle C(T)\rangle \left\{ \langle C(T)\rangle + \sum_{j=1}^{7} Y_{\infty,j}\left(1 - \frac{1-e^{-\alpha_j T}}{\alpha_j T}\right)\right\}. \end{aligned} \tag{10.94}$$

Based on the results derived above, the measurement principle of the Feynman-α method is explained below. When there is no fission chain reaction in a target system, the probability distribution of the randomly detected neutrons $C(T)$ follows a Poisson distribution. Then, the mean $\langle C(T)\rangle$ and the variance $\sigma^2(T)$ are equal to each other. On the other hand, as shown in Fig. 10.5, as the number of the fission chain reaction reactions in the target core increases, the statistical fluctuation of $C(T)$ becomes larger so that the variance $\sigma^2(T)$ is larger than the mean $\langle C(T)\rangle$. In the Feynman-α method, based on the statistical characteristics of the reactor noise described above, the relative difference between the variance and the mean in the measured neutron counts is quantified to investigate how much the probability distribution of $C(T)$ differs from the Poisson distribution. Namely, the neutron correlation coefficient factor $Y(T)$ is evaluated as follows:

$$Y(T) \equiv \frac{\sigma^2(T)}{\langle C(T)\rangle} - 1 = \frac{\langle C(T)(C(T)-1)\rangle - \langle C(T)\rangle^2}{\langle C(T)\rangle}, \tag{10.95}$$

where the final expression of Eq. (10.95) is derived by the following definition of the variance $\sigma^2(T)$ (i.e., the mean of the square $\langle C^2(T)\rangle$ minus the square of the mean $\langle C(T)\rangle^2$):

$$\sigma^2(T) = \left\langle (C(T) - \langle C(T)\rangle)^2 \right\rangle = \left\langle C^2(T)\right\rangle - \langle C(T)\rangle^2. \tag{10.96}$$

Finally, by substituting Eqs. (10.93) and (10.94) into Eq. (10.95), the theoretical formula for the neutron correlation factor $Y(T)$ can be obtained as follows:

$$Y(T) = \sum_{j=1}^{7} Y_{\infty,j}\left(1 - \frac{1-e^{-\alpha_j T}}{\alpha_j T}\right), \tag{10.97}$$

where $\alpha_j = -\omega_j$, and ω_j is the negative solution of the reactivity equation of Eq. (2.62) shown in Fig. 2.7 ($\omega_7 \ll \omega_6 < \cdots < \omega_2 < \omega_1$). Therefore, in the right-hand side of Eq. (10.97), the $j = 7$th term corresponds to the prompt neutron component, and the other terms of $1 \leq j \leq 6$ correspond to the delayed neutron components, respectively ($\alpha_7 \gg \alpha_6 > \cdots > \alpha_2 > \alpha_1$). By focusing on the time region where the contributions of the delayed neutron components are sufficiently small (e.g., $T \ll 1/\alpha_6 \approx 0.3$ s), the theoretical formula of Eq. (10.97) can be simplified as follows:

$$Y(T) \approx Y_{\infty,\mathrm{p}}\left(1 - \frac{1 - \mathrm{e}^{-\alpha T}}{\alpha T}\right) + c_1 T, \tag{10.98}$$

$$Y_{\infty,\mathrm{p}} \approx \varepsilon \frac{\langle \nu(\nu - 1)\rangle}{\langle \nu \rangle^2} \frac{1}{(\beta_{\mathrm{eff}} - \rho)^2}\left(1 + \frac{\langle \nu \rangle \langle q(q-1)\rangle}{\langle q \rangle \langle \nu(\nu - 1)\rangle}(-\rho)\right), \tag{10.99}$$

$$c_1 \approx \frac{1}{2}\sum_{j=1}^{6} Y_{\infty,j}\alpha_j, \tag{10.100}$$

where $Y_{\infty,\mathrm{p}}$ represents the saturation value for the prompt neutron component of the Y value, and c_1 is the correction term due to the delayed neutron components. In the theoretical formula of Eq. (10.99) for $Y_{\infty,\mathrm{p}}$, $\langle \nu(\nu - 1)\rangle / \langle \nu \rangle^2$ is the nuclear data known as the Diven factor [31], e.g., $\langle \nu(\nu - 1)\rangle / \langle \nu \rangle^2 = 0.796 \pm 0.003$ when the thermal fission reaction of ^{235}U is dominant [30]. Note that the underestimation term due to the dead time effect of the neutron detection process [32] is neglected in the theoretical derivation for Eq. (10.98).

Based on the theoretical formula derived above, the subcriticality can be obtained from the reactor noise measurement result with the bunching technique shown in Fig. 10.9. The procedure of the Feynman-α method is described below:

(1) As shown in Fig. 10.9, the time series data of neutron counts C_i are successively measured per a certain counting gate width T_0 ($1 \leq i \leq N$), where N is the total number of count data.
(2) The sample mean $\overline{C}$ and the unbiased sample variance s^2 are estimated from the measured time series data:

$$\overline{C} = \frac{1}{N}\sum_{i=1}^{N} C_i, \tag{10.101}$$

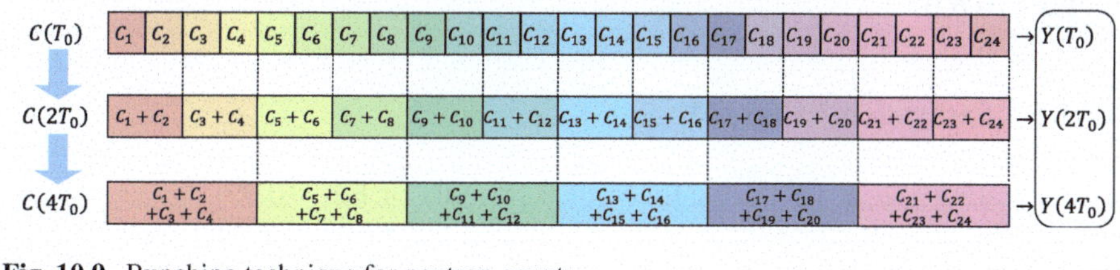

Fig. 10.9 Bunching technique for neutron counts

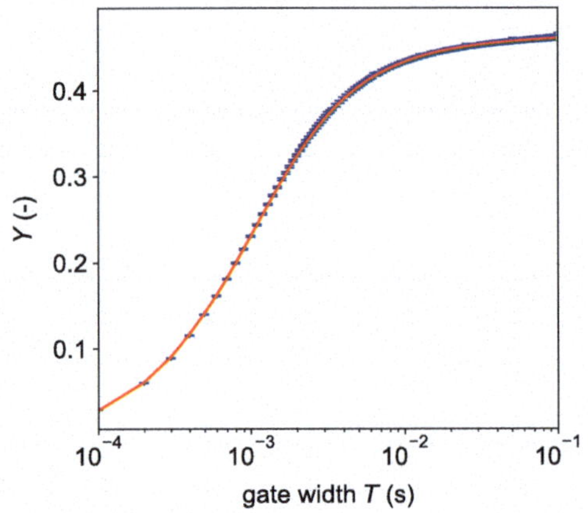

Fig. 10.10 Variation of Y with respect to counting gate width T

$$s^2 = \frac{1}{N-1}\sum_{i=1}^{N}(C_i - \overline{C})^2. \tag{10.102}$$

(3) Using $\overline{C}$ and s^2 as the mean $\langle C \rangle$ and the variance σ^2 in Eq. (10.95), the neutron correlation factor Y is evaluated.
(4) Using the bunching technique (i.e., by combining the adjacent neutron counts such as C_i and C_{i+1} as shown in Fig. 10.9), other time series data with different counting gate widths ($2T_0, 4T_0, \cdots$) can be obtained from the original time series data. Then, the corresponding Y values ($Y(2T_0), Y(4T_0), \cdots$) are evaluated to analyze the variation of $Y(T)$ with respect to T.
(5) By the nonlinear least-squares method using the fitting formula of Eq. (10.98) for the measured $Y(T)$ shown in Fig. 10.10, the prompt neutron decay constant α and the saturation value $Y_{\infty,\mathrm{p}}$ are obtained, respectively.
(6) If the point kinetics parameters (Λ and β_{eff}) are known, the measured α value can be converted into the subcriticality $-\rho$ based on Eq. (10.60).
(7) Alternatively, when the reactor noise measurements are carried out for multiple subcritical cores by changing control rod patterns while the external neutron source positions and strength are kept constant, the prompt neutron decay constant at the critical state, α_{c}, can be estimated. Then, the subcriticality in dollar units $-\rho/\beta_{\mathrm{eff}}$ can be obtained by the Simmons-King method of Eq. (10.62).

10.5.1.5 Saturation Value of Y

By substituting $-\rho = 0$ into the theoretical formula of Eq. (10.99), the saturation value for the prompt component of the Y value at the delayed critical state, $Y_{\infty,\mathrm{p,c}}$, can be derived as follows:

$$Y_{\infty,\mathrm{p,c}} = \varepsilon \frac{\langle \nu(\nu-1) \rangle}{\langle \nu \rangle^2} \frac{1}{\beta_{\mathrm{eff}}^2}. \tag{10.103}$$

As can be seen from Eq. (10.99), if the target core is a relatively shallow subcritical system ($|-\rho| \ll 1$) and/or the external neutron source simultaneously emits only one neutron per decay (e.g., Am-Be source, $\langle q(q-1) \rangle = 0$), the correlation term due to $\langle q(q-1) \rangle$ in the saturation value $Y_{\infty,\mathrm{p}}$ is negligible. Consequently, in a similar way to the Simmons-King method, the ratio of the saturation values at the critical state and the target subcritical state, $Y_{\infty,\mathrm{p,c}}$ and $Y_{\infty,\mathrm{p}}$, can be used to estimate the subcriticality in dollar units as follows:

$$\frac{Y_{\infty,\mathrm{p,c}}}{Y_{\infty,\mathrm{p}}} = \left(1 - \frac{\rho}{\beta_{\mathrm{eff}}}\right)^2. \tag{10.104}$$

In order to simplify the theoretical derivation process, this section assumes that the target system is an infinite homogeneous system so that the spatial effect on the Y value can be neglected. If the subcriticality of the target system is relatively shallow, the spatial effect due to the higher-order mode component becomes small. Then, it is possible to derive the theoretical formula for the Y value considering only the fundamental mode [33, 34]. Note that, if the spatial distribution of the fundamental mode can be approximated to be unchanged, the spatial correction term in the ratio of $Y_{\infty,\mathrm{p,c}}/Y_{\infty,\mathrm{p}}$ can be eliminated by taking the ratio of the fundamental mode components of the Y values. Consequently, when the fundamental mode approximation is applied to the ratio of $Y_{\infty,\mathrm{p,c}}/Y_{\infty,\mathrm{p}}$, the theoretical expression is equivalent to Eq. (10.104).

10.5.1.6 Statistical Uncertainty of Y

In an actual reactor noise measurement, the total number of count data N is finite due to the limited total measurement time. Consequently, the Y value based on the sample estimates by Eqs. (10.101) and (10.102) inevitably has the statistical uncertainty σ_Y. Recently, various methods have been proposed to quantify the statistical uncertainty σ_Y, e.g., (1) the analytical method based on the uncertainty propagation and (2) the resampling methods such as the moving block bootstrap method [35]. For example, if the subcriticality $-\rho$ in a target core is approximately less than $10\%\Delta k/k$, the statistical uncertainty σ_Y can be easily estimated using the following formula with the total number of count data N, the mean of neutron counts $\overline{C}$, and the neutron correlation factor Y [35]:

$$\sigma_Y \approx (Y+1)\sqrt{\frac{Y(2Y+1)(5Y+2)}{N(Y+1)^2\overline{C}} + \frac{2}{N-1}}. \tag{10.105}$$

10.5.2 Measurement Procedure

To carry out the reactor noise measurement, it is necessary to configure a steady-state subcritical core with an external neutron source. For example, a primary neutron source such as an Am-Be neutron source or a ^{252}Cf spontaneous neutron source is inserted into the target core as an external neutron source. Alternatively, in the case of uranium-aluminum (U-Al) alloy fuel previously used at KUCA, there is a weak inherent neutron source due to the (α,n) reaction of ^{27}Al induced by the α-decay of U in the nuclear fuel [36]. The weak inherent neutron source can also be used as an external neutron source for the reactor noise measurement, e.g., the long-time measurement under a shutdown condition [33–35].

The measurement of the reactor noise with high statistical precision requires a neutron detector (e.g., ^{3}He or BF_3 proportional counter) having as high detection efficiency as possible. Then, a signal processing circuit system should be configured to continuously measure the time series data of neutron counts. For example, a digital MCA as described in Sect. 9.2 can be used to obtain the list-mode data relating to the neutron detection time for each detector. The list-mode data measurement enables various reactor noise analysis methods such as the Feynman-α method, the Rossi-α method [10], and so on.

10.5.2.1 Histogram of Neutron Counts Following Poisson Distribution

As a preliminary experiment for a Poisson source that emits one neutron per decay, the procedure for measuring a histogram of neutron counts is described as follows:

(1) A neutron detector (^{3}He or BF_3 proportional counter, etc.) is placed near the storage of the Am-Be neutron source surrounded by paraffin. The signal processing circuit system is configured to carry out the reactor noise measurement so that the list-mode data of neutron detection time can be measured using the digital MCA.
(2) The reactor noise measurement is carried out over a total measurement time of approximately 100 s. The measured list mode data relating to neutron detection time is processed with a basic counting gate width T_0 and the total number of count data N to obtain the time series data of neutron counts $\vec{C} = (C_1, C_2, \cdots, C_N)$.
(3) By illustrating the histogram (or frequency distribution) of $\vec{C}$, the neutron correlation factor Y value is evaluated.

10.5.2.2 Reactor Noise Measurements in Subcritical Systems

For a target subcritical core with an external neutron source, the procedure for measuring the prompt neutron decay constant using the Feynman-α method is described as follows:

(1) A neutron detector (^{3}He or BF_3 proportional counter) is installed into the target subcritical core. The signal processing circuit system is configured to carry out the reactor noise measurement.
(2) To configure the target subcritical system, six control rods (three control rods and three safety rods) are inserted and/or withdrawn in an arbitrary pattern i.
(3) After sufficient time has passed for the source-driven subcritical system, where some kind of external neutron source exists, the neutron count rate is checked to confirm the steady state of the target system.
(4) The digital MCA is used for the list-mode measurement. By processing with a basic counting gate width T_0 and the total number of count data N, the time series data of neutron counts $\vec{C} = (C_1, C_2, \cdots, C_N)$ are obtained. Here, T_0 is determined according to the magnitude of the prompt neutron decay constant α (e.g., $T_0 = 0.1$ ms $= 10^{-4}$ s).
(5) The measured time series is input to a processing tool for the Feynman-α method prepared in advance, to evaluate the prompt neutron decay constant α and the saturation value $Y_{\infty,\text{p}}$.
(6) For the ith control rod pattern of the subcritical system with the same external neutron source, the following parameters are summarized: the count rate R_i, the prompt neutron decay constant α_i, and the saturation value $Y_{\infty,\text{p},i}$. In the same manner as the experiment of the approach-to-criticality, the inverse count rate R_0/R_i is estimated using the measured count rate R_i, where R_0 corresponds to the count rate measured at the deepest subcritical core. After plotting the measured α_i and $Y_{\infty,\text{p},i}$ with respect to R_0/R_i, the α_c and $Y_{\infty,\text{p,c}}$ values at the critical state are estimated by extrapolating α_i and $Y_{\infty,\text{p},i}$ at $R_0/R_i \to 0$.
(7) If the point kinetics parameters (Λ and β_eff) are known, the subcriticality $-\rho$ can be converted from the measurement result of α using Eq. (10.60). Alternatively, if α_c and $Y_{\infty,\text{p,c}}$ at the critical state can be estimated, the subcriticality in dollar units is obtained based on Eqs. (10.62) and (10.104).

10.5.3 Discussions

Let us discuss the measurement results of the Feynman-α method from the following viewpoints:

(1) From the time series data of neutron counts $\vec{C} = (C_1, C_2, \cdots, C_N)$ measured for a non-neutron multiplication system with an Am-Be neutron source, calculate the sample mean $\overline{C}$ and the unbiased sample variance s^2. Check whether the histogram of $\vec{C}$ follows a Poisson distribution $P_\lambda(C)$, where the population parameter (or the expectation of C) λ is set to $\lambda = \overline{C}$.
(2) In a similar manner to the discussion point (1), based on the time series data of neutron counts $\vec{C}$ measured for a source-driven subcritical system, check whether the histogram of $\vec{C}$ follows a Poisson distribution $P_\lambda(C)$. Especially, investigate how much the difference from the Poisson distribution $P_\lambda(C)$

changes when the counting gate width $T = kT_0$ is increased or decreased using the bunching techniques.

(3) Depending on the magnitude of the subcriticality $-\rho$, how does the variation of $Y(T)$ (or the magnitudes of the prompt neutron decay constant α and the saturation value $Y_{\infty,\mathrm{p}}$) change? Discuss the physical reasons for the above observations.

(4) For each subcritical system with the ith control rod pattern, evaluate the reference value of subcriticality $-\rho_{\mathrm{ref},i}$ based on the measurement results of the excess reactivity and the control rod worth. Then, compare the subcriticality $-\rho_i$ measured by the Feynman-α method with the reference value $-\rho_{\mathrm{ref},i}$, and discuss the main cause for the difference from $-\rho_{\mathrm{ref},i}$.

(5) As explained in Sect. 10.5.1.6, there is statistical uncertainty in the measured neutron correlation factor Y. How can the statistical uncertainty of the neutron correlation factor Y be reduced? For example, based on Eq. (10.105), discuss how to reduce the relative statistical uncertainty σ_Y/Y.

References

1. Tsuji M, Suzuki N, Shimazu Y (2003) Subcriticality measurement by neutron source multiplication method with a fundamental mode extraction. J Nucl Sci Technol 40:158–169. https://doi.org/10.1080/18811248.2003.9715346
2. Suzaki T (1991) Subcriticality determination of low-enriched UO_2 lattices in water by exponential experiment. J Nucl Sci Technol 28:1067–1077. https://doi.org/10.1080/18811248.1991.9731473
3. Sastre CA (1960) The measurement of reactivity. Nucl Sci Eng 8:443–447. https://doi.org/10.13182/NSE60-A25827
4. Sjöstrand NG (1956) Measurements on a subcritical reactor using a pulsed neutron source. Arkiv för Fysik 11:233–246
5. Simmons BE, King JS (1958) A pulsed neutron technique for reactivity determination. Nucl Scie Eng 3:595–608. https://doi.org/10.13182/NSE3-595-608
6. Hogan WS (1960) Negative-reactivity measurements. Nucl Sci Eng 8:518–522. https://doi.org/10.13182/NSE60-A25838
7. Williams MMR (1974) Random processes in nuclear reactors. Pergamon Press, Oxford, UK
8. Pázsit I, Pál L (2008) Neutron fluctuations: a treatise on the physics of branching processes. Elsevier, Oxford, UK
9. Feynman RP, de Hoffmann F, Serber R (1956) Dispersion of the neutron emission in U-235 fission. J Nucl Energy 3:64-IN10. https://doi.org/10.1016/0891-3919(56)90042-0
10. Orndoff JD (1957) Prompt neutron periods of metal critical assemblies. Nucl Sci Eng 2:450–460. https://doi.org/10.13182/NSE57-A25409
11. Shimazu Y (2014) A simple procedure to estimate reactivity with good noise filtering characteristics. Ann Nucl Energy 73:392–397. https://doi.org/10.1016/j.anucene.2014.07.005
12. Keepin GR (1965) Physics of nuclear kinetics. Addison-Wesley, Reading, MA, USA
13. Arce GR (2004) Nonlinear signal processing: a statistical approach. Wiley-Interscience, Hoboken, NJ, USA
14. Hoogenboom JE, van der Sluijs AR (1988) Neutron source strength determination for on-line reactivity measurements. Ann Nucl Energy 15:553–559. https://doi.org/10.1016/0306-4549(88)90059-X

15. Nishioka F, Endo T, Yamamoto A et al (2022) Applicability of dynamic mode decomposition to estimate fundamental mode component of prompt neutron decay constant from experimental data. Nucl Sci Eng 196:133–143. https://doi.org/10.1080/00295639.2021.1968225
16. Endo T, Nishioka F, Yamamoto A et al (2022) Area ratio method using dynamic mode decomposition. In: Proceeding PHYSOR 2022, Pittsburgh, PA, USA, 15–20 May 2022
17. Gozani T (1962) A modified procedure for the evaluation of pulsed source experiments in subcritical reactors. Nukleonik 4:348–349
18. Katano R, Yamanaka M, Pyeon CH (2019) Application of linear combination method to pulsed neutron source measurement at Kyoto University Critical Assembly. Nucl Sci Eng 193:1394–1402. https://doi.org/10.1080/00295639.2019.1624084
19. Katano R, Yamanaka M, Pyeon CH (2020) Measurement of prompt neutron decay constant with spallation neutrons at Kyoto university critical assembly using linear combination method. J Nucl Sci Technol 57:169–176. https://doi.org/10.1080/00223131.2019.1671911
20. Endo T, Nishioka F, Yamamoto A et al (2022) Application of dynamic mode decomposition to Rossi-α method in a critical state using file-by-file moving block bootstrap method. J Nucl Sci Technol 59:1117–1126. https://doi.org/10.1080/00223131.2022.2030260
21. Pyeon CH (ed) (2021) Accelerator-driven system at Kyoto university critical assembly, Springer Singapore, Singapore
22. Planche T, Lagrange J-B, Yamakawa E et al (2011) Harmonic number jump acceleration of muon beams in zero-chromatic FFAG rings. Nucl Instrum Meth Phys Res A 632:7–17. https://doi.org/10.1016/j.nima.2010.12.190
23. Lagrange J-B, Planche T, Yamakawa E et al (2012) Straight scaling FFAG beam line. Nucl Instrum Methods Phys Res A 691:55–63. https://doi.org/10.1016/j.nima.2012.06.058
24. Yamakawa E, Uesugi T, Lagrange J-B et al (2013) Serpentine acceleration in zero-chromatic FFAG accelerators. Nucl Instrum Meth Phys Res A 716:46–53. https://doi.org/10.1016/j.nima.2013.03.061
25. Pyeon CH, Yamanaka M, Oizumi A et al (2019) First nuclear transmutation of ^{237}Np and ^{241}Am by accelerator-driven system at Kyoto University Critical Assembly. J Nucl Sci Technol 56:684–689. https://doi.org/10.1080/00223131.2019.1618406
26. Courant ED, Wallace PR (1947) Fluctuations of the number of neutrons in a pile. Phys Rev 72:1038–1048. https://doi.org/10.1103/PhysRev.72.1038
27. Pál L (1958) On the theory of stochastic processes in nuclear reactors. Nuovo Cim 7:25–42
28. Bell GI (1965) On the stochastic theory of neutron transport. Nucl Sci Eng 21:390–401. https://doi.org/10.13182/NSE65-1
29. Yamane Y, Pázsit I (1998) Heuristic derivation of Rossi-alpha formula with delayed neutrons and correlated source. Ann Nucl Energy 25:1373–1382. https://doi.org/10.1016/S0306-4549(98)00017-6
30. Holden NE, Zucker MS (1988) Prompt neutron emission multiplicity distribution and average values (nubar) at 2200 m/s for the fissile nuclides. Nucl Sci Eng 98:174–181. https://doi.org/10.13182/NSE88-A28498
31. Diven BC, Martin HC, Taschek RF, Terrell J (1956) Multiplicities of fission neutrons. Phys Rev 101:1012–1015. https://doi.org/10.1103/PhysRev.101.1012
32. Yamane Y, Ito D (1996) Feynman-α formula with dead time effect for a symmetric coupled-core system. Ann Nucl Energy 23:981–987. https://doi.org/10.1016/0306-4549(95)00083-6
33. Endo T, Yamamoto A, Yamanaka M, Pyeon CH (2019) Experimental validation of unique combination numbers for third- and fourth-order neutron correlation factors of zero-power reactor noise. J Nucl Sci Technol 56:322–336. https://doi.org/10.1080/00223131.2019.1580625
34. Endo T, Nishioka F, Yamamoto A et al (2023) Theoretical derivation of a unique combination number hidden in the higher-order neutron correlation factors using the Pál-Bell equation. Nucl Sci Eng 197:176–188. https://doi.org/10.1080/00295639.2022.2049992

35. Endo T, Yamamoto A (2019) Comparison of theoretical formulae and bootstrap method for statistical error estimation of Feynman-α method. Ann Nucl Energy 124:606–615. https://doi.org/10.1016/j.anucene.2018.10.032
36. Shiozawa T, Endo T, Yamamoto A et al (2014) Investigation on subcriticality measurement using inherent neutron source in nuclear fuel. In: Proceeding PHYSOR2014, Kyoto, Japan, 28 Sep–3 Oct, 2014

Appendix
Discussions

A.1 Approach-to-Criticality Experiment

The report of this experiment should be clearly written and should be easy to understand for readers who have possibility to use the measurement data obtained through the experiment. It is desirable that the readers can reproduce the measurement results with the same procedure and they can understand which conclusions can be drawn from the results. Based on this consideration, the following should be summarized as measurement results:

- Locations of the fuel frames, the control rods, the safety rods, the neutron detectors, and the external neutron source in the reference core.
- Fuel loading information in the reference core and the other cores.
- Prediction of the number of the fuel plates to achieve the criticality at each measurement step.
- Explanations of how to determine the number of the added fuel plates.
- Table of data of the neutron count rates measured by the experimental group participants are belonging to.
- Figure of the inverse count rates for each neutron detector.
- Figure presenting the relation between the water level of the core tank and the neutron count rate of the detector.
- Operation log when a critical state is attained.
- Measurement results of the excess reactivity.
- Minimum number of fuel plates to attain the criticality. After the control rod calibration experiment, the reactivity worth of one fuel plate α_{plate} is derived, and the minimum number of fuel plates to achieve the criticality is calculated.

C. H. Pyeon et al., *Reactor Laboratory Experiments at Kyoto University Critical Assembly*, https://doi.org/10.1007/978-981-97-8070-9

A.1.1 Inverse Count Rate

It is expected that various curves of the inverse count rates dependent on the detector positions are obtained. Discuss this dependence of the shape of the curves on the detector positions by considering the relation in spatial positions among a reactor core, detectors, external source, and the direction of the fuel plate addition. It is also beneficial to compare the curves with those obtained numerically in the preliminary exercises described in Sect. 3.4.2.

A.1.2 Light Water Reflector Thickness Equivalent to the Infinite Thickness

Let us remember the change in the neutron count rate when decreasing the water level of the reactor tank from the fully-flooded state through the measurement in Sect. 3.2.1 (5). This change in the neutron count rate is caused by the change in the thickness of the top reflectors; reduction of the reflector thickness reduces the capability of neutron reflection of the water around the top of the reactor core.

Some of neutrons leaked from the reactor core are returned by the reflector and again contribute to the fission chain reaction. When the reflector thickness is increased from zero, this effect is expected to be enhanced. Qualitatively, if the reflector thickness becomes a certain value, the probability of leaked neutrons to reach the reflector with this thickness becomes negligible, and the neutron reflection capability would be saturated. The reflector having this thickness can be regarded as the infinite-thickness reflector. If such phenomena are observed during this experiment, estimate the reflector thickness which is equivalent to the infinite thickness. Furthermore, discuss whether the qualitative explanations on this are possible or not by considering the spatial shapes of neutron flux obtained by the preliminary exercises.

A.1.3 The Number of Fuel Plates to Achieve Criticality

From the difference in the number of the fuel plates to attain the criticality between the predicted value in the preliminary exercises and the actual value obtained through the experiment ΔN, calculate the difference in the unit of reactivity $\Delta\rho$. It can be obtained as

$$\Delta\rho = \Delta N \times \alpha_{\text{plate}}, \tag{3.12}$$

where α_{plate} is the reactivity worth of one fuel plate.

The number of the fuel plates to achieve the criticality predicted from the curves of the inverse count rate is different among the control rod insertion patterns $j = 1, 2,$

and 3. This difference in the prediction comes from the difference in the control rod insertion pattern. For example, the prediction at $j = 2$ is for the reactor core in which one control rod is fully inserted and the other two are fully withdrawn, and that at $j = 3$ is for the core in which all these three control rods are fully withdrawn. This difference in the prediction can be explained by the difference in k_{eff} dependent on the control rod insertion pattern. Convert the difference in the predicted number of the fuel plates to achieve the criticality between $j = 1$ or 2 and $j = 3$ to the reactivity by using α_{plate}. Furthermore, compare these reactivities with those obtained through the actual measurement in the control rod calibration experiment considering the detector positions.

A.2 Control Rod Calibration Experiment

Items for the exercise are as follows:

(1) In the measurement of the stable period method, derive the reactivity ρ by hands from the stable period T measured with stopwatches and Eq. (4.3), and summarize the results with the table in which the contributions of the group-wise delayed neutrons to the reactivity are clearly described. Please ask instructors which measurement is concerned.
(2) Obtain the control rod reactivity worth from the measurement results with the stable period method and the rod drop method (the extrapolation method and the integral method).
(3) Obtain the integral and differential reactivity curves for the control rod to which the measurement is carried out with the stable period method.
(4) Derive the shutdown margin.
(5) Derive the excess reactivity.
(6) Derive the reactivity worth per one fuel plate from the result in the exercise (5) and the excess reactivity of the reactor core in the approach-to-criticality experiment.

The items for discussions are as follows:

(1) Based on the results of the exercise (2), discuss the difference in the reactivity worth among different control rods qualitatively and quantitatively. The control rod reactivity worth derived from the assumption of the cosine-shaped neutron flux distribution discussed in Sect. 4.6 would be useful if it is applied to the radial x–y plane of the core.
(2) Compare the reactivity worth deduced from the stable period method with that from the rod drop method. If there is a difference, discuss its reason.
(3) Compare the reactivity worth deduced from the integral method with that from the extrapolation method. If there is a difference, discuss its reason.
(4) Let $\rho(H)$ be the control rod worth deduced from the stable period method and plot a semi-analytical control rod (integral) worth curve using Eq. (4.20)

in Sect. 4.5 with the measurement-based results obtained in the exercise (3). Please be careful that the measurement results are on the "withdrawal length" and the results with Eq. (4.20) are on the "insertion length".
(5) Compare the two curves obtained in the item (4) with each other, and if there is a difference, discuss the reason. Please be careful on the following; Eq. (4.20) is derived for a bare reactor in which neutron flux spatial distribution is approximated by the sine/cosine function, so the parameters H and x should be carefully examined. Results obtained with the experiment for the neutron flux spatial distribution would be also useful. The axial geometric configurations of the control rod and the fuel plate in the KUCA-C core can be found in Fig. 4.3.
(6) Let us assume that the C2 control rod is calibrated by the stable period method and the C3 control rod is used for the reactivity compensation. When the C2 reactivity worth is slightly larger than the C3 reactivity worth, the reactor core does not reach a critical state when C2 is fully withdrawn and C3 is fully inserted. Describe how to terminate the C3 calibration over an entire stroke in such a case.
(7) Control rod calibration is generally performed with the positive stable period, and the negative stable period is generally not used. Discuss the reason.

A.3 Measurements of Reaction Rates

A.3.1 Experimental Reports

(1) Purpose of experiment.
(2) Experimental procedures.
(3) Experimental settings.

- Name of core, core configuration and irradiation positions of Au wires (refer to relevant documents),
- Operation number (Run No.) during irradiation, positions of control rods and safety rods, water level, core temperature, indicated values of linear power meter and γ-ray area monitor, start time, and end time of Au wire irradiation,
- Type of γ-ray detector used to measure radioactivity.

A.3.2 Reaction Rate Distributions by Gold Wires

(1) Measured data of radioactivity

For the results of the radioactivity measurements of Au wires, consider the following items for each of bare and Cd-tubed Au wires:

- Number of Au wires,
- Locations of Au wires in the core,

- Start time of radioactivity measurement,
- Irradiation time,
- Waiting time,
- Measurement time,
- Counts and count rates for the γ-ray at 411.8 keV of all Au wire samples obtained by cutting,
- Background counts and count rates without any Au wire sample,
- Net count rates of the γ-ray at 411.8 keV (γ-ray count rates—background count rates) of all Au wire samples,
- Masses of all Au wire samples,
- Relative value of activation reaction rate shown in Eq. (5.17) of all Au wire samples.

(2) Discussions

- Compare the reaction rate distributions by the bare Au wires in directions x, y, and z (Fig. 1.9 in Chap. 1), and discuss the differences in shape and the peaks in the reflector regions relative to the core center.
- Describe reasons of the difference between the shapes of reaction rate distributions by bare and Cd-tubed Au wires in the direction z.
- Using the reaction rate distributions by bare and Cd-tubed Au wires in the direction z, show the relative spatial distributions of reaction rates corresponding to total and epi-thermal neutron fluxes in the fuel region in the direction z. Also, approximate the activation reaction rate corresponding to thermal neutron flux from the difference between the two reaction rate distributions.
- Using the reaction rate distributions by bare and Cd-tubed Au wires in the direction z, determine the Cd ratio of reaction rate distributions at the fuel region in the direction z with the use of Eq. (5.18). Also, from the viewpoint of neutron spectrum, discuss the differences in positions using the Cd ratio results in fuel and light-water reflector regions.

A.4 Determination of Neutron Flux

A.4.1 Experimental Reports

(1) Purpose of experiment.
(2) Experimental procedures.
(3) Experimental settings.

- Name of core, core layout and irradiation position of Au wire (refer to relevant documents),

- Operation number (Run No.) during irradiation, positions of control rods and safety rods, water level, core temperature, indicated values of linear power meter and γ-ray area monitor, start time, and end time of Au wire irradiation,
- Type of γ-ray detector used to measure radioactivity.

A.4.2 Determination of Thermal Neutron Flux and Reactor Power Using Gold Foils

(1) Measured data of radioactivity

For the results of the radioactivity measurements of Au foils, obtain the following items for each of the bare and Cd-covered Au foils:

- Locations of Au foils in the core,
- Start time of radioactivity measurement,
- Irradiation time,
- Waiting time,
- Measurement time,
- Counts and count rates for the γ-ray at 411.8 keV,
- Background counts and count rates,
- Net count rates of the γ-ray at 411.8 keV (γ-ray count rates - background count rates),
- Data on radioactivity measurements using standard sources,
- Detection efficiency of the γ-ray at 411.8 keV,
- Saturated activity in Eq. (5.15),
- Mass of Au foil samples,
- Reaction rate per mass in Eq. (5.16).

(2) Discussions

- Based on the results of the radioactivity measurements of Au foils, determine the absolute values of thermal neutron flux and reactor power at the irradiation location according to the item (3) in Sect. 6.7.2. Also include the Cd ratio obtained by the Au foils and the various other factors used in the correction.
- Using the results of the relative reaction rate distributions from the Au wires in directions x and z obtained in Chap. 5, evaluate quantitatively the relationship between the irradiation position and the center of the fuel region with respect to the reaction rates, and determine the absolute values of thermal neutron flux and reactor power at the core center (centers in directions x, y, and z).

A.5 Exponential Experiments with Natural Uranium

A.5.1 *Experimental Reports*

(1) From the results of steps (4) through (9) in Sect. 7.6.1, obtain the reaction rate distributions measured from ^{3}He#1 and ^{3}He#2 at the number of fuel assemblies, 10, 14, 19, and 24. Using the results, the findings are the spatial decay constants $\mu_{10}^{\mathrm{He-3\#1,\#2}}$, $\mu_{14}^{\mathrm{He-3\#1,\#2}}$, $\mu_{19}^{\mathrm{He-3\#1,\#2}}$, and $\mu_{24}^{\mathrm{He-3\#1,\#2}}$ for 10, 14, 19, and 24 fuel assemblies, respectively, by the least-squares method. In terms of this, make discussions about the relationship between the number of fuel assemblies and the spatial decay constant.

(2) Using the results obtained from step (10) in Sect. 7.6.1, the findings are the inverse multiplication curves by ^{3}He#1 and ^{3}He#2. Here, make discussions about the shape of the curves obtained.

(3) When the count rate C_i, detection efficiency ε, neutron source strength S, and multiplication factor $k_{\mathrm{s,i}}$ of the system are obtained by steps (4)–(9) in Sect. 7.6.1, Eq. (7.43) is approximated by the neutron source multiplication method (Chap. 10) as follows:

$$C_i = \frac{\varepsilon S}{1 - k_{s,i}}. \tag{7.43}$$

The detection efficiency ε and neutron source strength S are then assumed to be constant. Also, assuming that the multiplication factor $k_{\mathrm{s,3}}$ for the system with 24 fuel assemblies ($i = 3$) is 0.50, obtain the multiplication factors $k_{\mathrm{s,0}}$, $k_{\mathrm{s,1}}$ and $k_{\mathrm{s,2}}$ for systems with 10 ($i = 0$), 14 ($i = 1$), and 19 ($i = 2$) fuel rods, using the count rate C_l

(4) The values of $k_{\mathrm{s,0}}$, $k_{\mathrm{s,1}}$, and $k_{\mathrm{s,2}}$ are obtained from two neutron detectors (^{3}He#1 and ^{3}He#2), and make discussions about the difference between the results of ^{3}He#1 and ^{3}He#2.

(5) Using the spatial decay constants obtained from the results of inverse multiplication and reaction rate distribution (steps (1) and (2) in Sect. 7.6.3), $\mu_{24}^{\mathrm{He-3\#1}}$, $\mu_{24}^{\mathrm{He-3\#2}}$, and μ_{24}^{In} for 24 fuel assemblies, make discussions about the differences between the results of spatial decay constants obtained from the ^{3}He detectors (#1 and #2) and the In wire.

A.5.2 *Reactor Physics Discussion*

(1) Using the spatial decay constant obtained from the In reaction rate distribution for 24 fuel assemblies (step (2) in Sect. 7.6.3) μ_{24}^{In}, obtain the material buckling B_m^2 from Eq. (7.16).

(2) For the In reaction rate distribution corresponding to the thermal neutron flux distribution ($\phi_{24}^{\mathrm{In}}(z)$) in the estimation of the spatial decay constant (step (2) in Sect. 7.6.3), obtain the diffusion distance L_{24}^{In} by the In wire using Eq. (7.27) by the least-squares method.
(3) Giving the material buckling B_m^2 and diffusion distance L_{24}^{In} obtained in discussions (1) and (2) in Sect. 7.7.2, respectively and the Fermi age τ (see Table 7. 1), the infinite multiplication factor k_∞ can then be obtained. Make discussions about the criticality of the experimental system made with NU fuel and polyethylene reflectors.
(4) When the system is in a critical steady state, using the diffusion equation with one-energy group, the material buckling B_m^2 can be expressed as follows:

$$B_m^2 = \frac{\nu\sum_{\mathrm{f}} - \sum_{\mathrm{a}}}{D}, \tag{7.44}$$

where ν is the average number of neutrons produced per fission, $\sum_{\mathrm{f}}$ the fission cross sections, and $\sum_{\mathrm{a}}$ the absorption cross sections. Here, discuss the possibility that the system is reached at a critical state, considering the composition of NU fuel and polyethylene moderator in the fuel assembly shown in Fig. 7.3.

A.6 Measurement of Uranium Enrichment

(1) Compare the results of $EN_{\mathrm{Case\,A}}$ and $EN_{\mathrm{Case\,B}}$ obtained by steps (2) and (3) in Sect. 8.4.4 with the theoretical enrichment (Case A: 46.6 wt%; Case B: 32.1 wt%) calculated from the combination of Cases A and B. Moreover, when observing a discrepancy between the results of experiments and references, discuss the reason of the discrepancy.
(2) Perform the γ-ray spectrum measurements on ^{232}Th samples using a NaI(Tl) detector (or HPGe detector) and compare them to the results of γ-ray spectrum measurements on U samples, and discuss any differences (this is an optional piece of homework).

A.7 Neutron Measurements

A.7.1 Observation of Signals from Neutron Detectors and Signal Processing Devices

(1) Compare the output signals from the preamplifier the shaping amplifier, and observe and measure the time width of each signal pulse. Describe the characteristics of each signal.

(2) Compare the output of a photomultiplier tube terminated in 50 Ω and that terminated in 1 MΩ, and describe the characteristics of each signal.
(3) Record how the time width of the signal pulse and the pulse height change when the time constant of the shaping amplifier is changed, and describe the characteristics.
(4) Compare the preamplifier output signals among various neutron detectors and describe the differences.

A.7.2 Measurement of Signal Pulse Height Spectra in Various Neutron Detectors

(1) Compare the signal pulse height spectra obtained from various neutron detectors when irradiated by thermal neutrons, and describe their characteristics, focusing on the differences between them.
(2) Compare the signal pulse height spectra obtained from various neutron detectors when irradiated by gamma rays, and describe their characteristics, focusing on the differences between them.
(3) Based on the signal pulse height spectra obtained when irradiated by thermal neutrons and gamma rays, set the signal discrimination level. The signals with pulse heights above this level is considered to be neutron signal. And state the criteria for determining the level.
(4) If possible, change the voltage applied to the detector, observe how the signal pulse height spectrum changes, and discuss the reasons.
(5) If you have more time, analyze the list-mode data and create the signal pulse height spectra.

A.7.3 Observation of Effect of Moderator

(1) Change the moderator geometry, record the change in the neutron signal count rate, compare how the response of each neutron detector changes, and discuss the reasons.
(2) Simulate the moderator geometry and neutron detectors using the Monte Carlo simulation code PHITS, and calculate the detector reaction rates. Compare the obtained results with the experimental results and discuss the results.

A.8 Subcriticality Measurements

A.8.1 Neutron Source Multiplication Method

(1) For comparison, estimate the reference value of the subcriticality in the targe core, $-\rho_{\text{target}}$, based on the control rod pattern with the known values of the excess reactivity and the control rod worth, which were previously measured in advance.
(2) Comparing $-\rho_{\text{target}}$ estimated by the neutron source multiplication method with the reference value obtained in the previous discussion (1), investigate the measurement accuracy (or the relative difference from the reference value) to quantitatively discuss the agreement of the neutron source multiplication method. If there is a significant difference between these results, discuss the main cause of the difference based on the approximation conditions assumed in the principle of the neutron source multiplication method.
    - Based on Eq. (10.15), discuss whether the subcriticality of the reference core can be considered as a near-critical system.
    - Discuss the differences in the measurement results using some detectors installed at the different positions, by focusing on the positional relation among the detector, the external neutron source, and the fuel region.
(3) Discuss whether there are differences in the measured subcriticality values of $-\rho_{\text{target}}$ between the neutron source multiplication method (the static method) and other measurement methods (e.g., the kinetic method and the reactor noise analysis method as explained in Sects. 10.3 through 10.5).
(4) Based on the measured neutron count rates in the approach-to-criticality experiment as explained in Chap. 3, try to estimate the subcriticality value for each subcritical state using the neutron source multiplication method.

A.8.2 Inverse Kinetics Method

(1) Discuss the relationship between the time dependence of the measured count rate $R(t)$ (e.g., magnitude and sign for the slope) and the reactivity $\rho(t)$ using the inverse kinetics method. For example, how is $\rho(t)$ inversely estimated for the following change of $R(t)$?
    (a) $R(t)$ is constant.
    (b) $R(t)$ increases with time.
    (c) $R(t)$ decreases with time.
(2) Discuss how to reduce the statistical uncertainty of the reactivity $\rho(t)$ using the inverse kinetics method.

(3) What is the difference in the estimated reactivity $\rho(t)$ between the inverse kinetics method and the neutron source multiplication method described in Sect. 10.2? For example, discuss how the differences in $\rho(t)$ between these methods are (a) during the transient state of the count rate due to the control rod operation and (b) at steady state after sufficient time has elapsed following the control rod operation.

(4) By carrying out a transient experiment to which the least-squares inverse kinetics method is applicable, estimate the effective external neutron source strength Q and the reactivity in dollar units $\rho_{\text{after}}/\beta_{\text{eff}}$ after the transient, based on Eq. (10.41). For these results of Q and $\rho_{\text{after}}/\beta_{\text{eff}}$ estimated by the least-squares inverse kinetics method, compare them with the source strength Q calibrated by Eq. (10.38) using the reference subcriticality (e.g., ρ_{ref} evaluated by the excess reactivity and the control rod worth) and the reactivity estimated by the inverse kinetics method with such calibrated Q. Then, discuss the reasons for the differences between them, if any.

A.8.3 *Pulsed Neutron Source Method*

(1) Evaluate the reference value of subcriticality $-\rho_{\text{ref},i}$ for the ith subcritical core, based on the experimental results of the excess reactivity and the control rod worth. Make a scatter plot of the measured α_i against $-\rho_{\text{ref},i}$. Then, estimate the prompt neutron decay constant at criticality α_c by extrapolating the α value at the critical state (or $-\rho = 0$).

- Or estimate α_c using the area ratio $\left(A_p/A_d\right)_i$ as the subcriticality in dollar units in Eq. (10.61).

(2) Check how well the subcriticality obtained by (i) the Simmons-King method of Eq. (10.62) agrees with that of (ii) the area ratio method (or the extrapolated area ratio method). If there is a difference between these methods, discuss the cause of the difference by changing the masking time in the fitting process of the pulsed neutron method.

(3) Calculate the reactivity ρ by substituting $-\alpha$ for ω (i.e., $\omega = -\alpha$) in the reactivity equation of Eq. (2.62) using the point kinetics parameters (prompt neutron lifetime ℓ or neutron generation time Λ and effective delayed neutron fraction β_{eff}) and the delayed neutron parameters. Then, compare the calculated ρ with the subcriticality estimated by the Simmons-King method of Eq. (10.62). Through this comparison, discuss the relationship between the prompt neutron decay constant α and the solution for ω_j of Eq. (2.62).

(4) Based on (i) the spatial relationship between the core, the pulsed neutron source, and the neutron detector and (ii) the magnitude of the subcriticality of the target core, investigate whether the experimental results of the pulsed neutron source method (time variation of the count rate $R_\tau(t)$, prompt neutron decay constant

α, and the area ratio A_p/A_d) differ depending on the neutron detector position. Then, discuss the reasons for these differences.

- By comparing the measured subcriticality with the reference value $-\rho_{ref}$ based on the excess reactivity and the control rod worth, discuss what kind of the condition for (i) the detector position and (ii) the magnitude of the subcriticality is suitable for applying the fundamental mode approximation.

A.8.4 Feynman-α Method

(1) From the time series data of neutron counts $\vec{C} = (C_1, C_2, \cdots, C_N)$ measured for a non-neutron multiplication system with an Am-Be neutron source, calculate the sample mean $\overline{C}$ and the unbiased sample variance s^2. Check whether the histogram of $\vec{C}$ follows a Poisson distribution $P_\lambda(C)$, , where the population parameter (or the expectation of C) λ is set to $\lambda = \overline{C}$.
(2) In a similar manner to the discussion point (1), based on the time series data of neutron counts $\vec{C}$ measured for a source-driven subcritical system, check whether the histogram of $\vec{C}$ follows a Poisson distribution $P_\lambda(C)$. Especially, investigate how much the difference from the Poisson distribution $P_\lambda(C)$ changes when the counting gate width $T = kT_0$ is increased or decreased using the bunching techniques.
(3) Depending on the magnitude of the subcriticality $-\rho$, how does the variation of $Y(T)$ (or the magnitudes of the prompt neutron decay constant α and the saturation value $Y_{\infty,p}$) change? Discuss the physical reasons for the above observations.
(4) For each subcritical system with the ith control rod pattern, evaluate the reference value of subcriticality $-\rho_{ref,i}$ based on the measurement results of the excess reactivity and the control rod worth. Then, compare the subcriticality $-\rho_i$ measured by the Feynman-α method with the reference value $-\rho_{ref,i}$, and discuss the main cause for the difference from $-\rho_{ref,i}$.
(5) As explained in Sect. 10.5.1.6, there is statistical uncertainty in the measured neutron correlation factor Y. How can the statistical uncertainty of the neutron correlation factor Y be reduced? For example, based on Eq. (10.105), discuss how to reduce the relative statistical uncertainty σ_Y/Y.

Index

C. H. Pyeon et al., *Reactor Laboratory Experiments at Kyoto University Critical Assembly*, https://doi.org/10.1007/978-981-97-8070-9

SPRINGER NATURE

GPSR Compliance

The European Union's (EU) General Product Safety Regulation (GPSR) is a set of rules that requires consumer products to be safe and our obligations to ensure this.

If you have any concerns about our products, you can contact us on ProductSafety@springernature.com

In case Publisher is established outside the EU, the EU authorized representative is:

Springer Nature Customer Service Center GmbH
Europaplatz 3
69115 Heidelberg, Germany